THREE BIO-REALMS

Biotechnology and the Governance of Food, Health, and Life in Canada

Biotechnology has become one of the most important issues in public policy and governance, altering the boundaries between the public and the private, the economic and the social, and further complicating the divide between what is scientifically possible and ethically preferred. Given the importance of biotechnology in shaping relations between the state, science, the economy, and the citizenry, a book that explores the Canadian biotechnology regime and its place in our democracy is timelier than ever.

Three Bio-Realms provides the first integrated examination of the thirty-year story of the democratic governance of biotechnology in Canada. G. Bruce Doern and Michael J. Prince, two recognized specialists in governance innovation and social policy, look at particular 'network-based' factors that seek to promote and to regulate biotechnology inside the state as well as at broader levels. Unmatched by any other book in its historical scope and range, *Three Bio-Realms* is sure to be read for years to come.

(Studies in Comparative Political Economy and Public Policy)

G. BRUCE DOERN is Distinguished Research Professor in the School of Public Policy and Administration at Carleton University and an emeritus professor in the Department of Politics at the University of Exeter.

MICHAEL J. PRINCE is Lansdowne Professor of Social Policy in the Faculty of Human and Social Development at the University of Victoria.

Studies in Comparative Political Economy and Public Policy

Editors: MICHAEL HOWLETT, DAVID LAYCOCK (Simon Fraser University), and STEPHEN MCBRIDE (McMaster University)

Studies in Comparative Political Economy and Public Policy is designed to showcase innovative approaches to political economy and public policy from a comparative perspective. While originating in Canada, the series will provide attractive offerings to a wide international audience, featuring studies with local, subnational, cross-national, and international empirical bases and theoretical frameworks.

Editorial Advisory Board

For a list of books published in the series, see page 267.

Three Bio-Realms

Biotechnology and the Governance
of Food, Health, and Life in Canada

G. BRUCE DOERN AND MICHAEL J. PRINCE

UNIVERSITY OF TORONTO PRESS
Toronto Buffalo London

© University of Toronto Press 2012
Toronto Buffalo London
www.utppublishing.com
Printed in Canada

ISBN 978-1-4426-4277-5 (cloth)
ISBN 978-1-4426-1154-2 (paper)

Library and Archives Canada Cataloguing in Publication

Doern, G. Bruce, 1942–
Three bio-realms : biotechnology and the governance of food,
health, and life in Canada / G. Bruce Doern and Michael J. Prince.

(Studies in comparative political economy and public policy)
Includes bibliographical references and index.
ISBN 978-1-4426-4277-5 (bound). – ISBN 978-1-4426-1154-2 (pbk.)

1. Biotechnology – Government policy – Canada – History.
2. Biotechnology – Government policy – Canada.   3. Biotechnology –
Canada.   I. Prince, Michael John, 1952 –   II. Title.   III. Series:
Studies in comparative political economy and public policy

TP248.195.C3D63 2012   338.4'766060971   C2012-902214-4

University of Toronto Press acknowledges the financial assistance to its
publishing program of the Canada Council for the Arts and the Ontario
Arts Council.

 Canada Council   Conseil des Arts
for the Arts   du Canada   ONTARIO ARTS COUNCIL
CONSEIL DES ARTS DE L'ONTARIO

University of Toronto Press acknowledges the financial support of the
Government of Canada through the Canada Book Fund for its publishing
activities.

Contents

Preface

Three Bio-Realms is about biotechnology as an enabling and transformative science in food, health, and human life, and its regulatory governance in Canada set within an international context. In addition, we have a good deal to say about relations of private and personal power, notions of democracy, and beliefs about nature, being human, and the self. In total, we treat the subject of biotechnology in a broad fashion, giving consideration not only to scientific research, business products and intellectual property, government policy and administration, but also to the related issues of family formation, sexual identity, food quality and safety, personalized medicine, and the challenges facing individuals and public health-care systems across the country.

We contend that biotechnologies alter our notions of food and nature, health and illness, and human life. In this sense, we observe a 'geneticization' of politics and public discourse along with a politicization of the life sciences. We further suggest that existing governance processes in Canada, while valuable, are too often politically and socially marginal. Larger and permanent arenas linked more closely to both parliamentary government and federalist democracy are needed, since bioethics will become ever more present and complex in the governance of biotechnology.

This book is the product of the authors' individual and collaborative work on biotechnology governance and its related aspects over the last decade. During this extensive period of reading, discussion, and interviews, we owe numerous debts of thanks, gratitude, and learning to many individuals and to many agencies and institutions involved directly and indirectly with the biotechnology governance story.

These individuals include in particular Roy Atkinson, Brian Colton, Derek Ireland, Jeff Kinder, Joseph Konvitz, Doug Kuntze, William Leiss,

Nancy Miller Chenier, Joan Murphy, George Redling, Gilles Rheaume, Marc Saner, Jac van Beek, Kernaghan Webb, and Anne Wiles.

We have benefited greatly from the work of the two peer review assessors for the University of Toronto Press who provided construc- tive and perceptive comments on an earlier draft of the manuscript. They have without doubt helped to make this a better book both con- ceptually and empirically.

At an agency and institutional level, thanks are owed to staff and advisors at the former Canadian Biotechnology Advisory Committee and the Canadian Biotechnology Secretariat, and at Industry Canada, BIOTECanada, the Canadian Food Inspection Agency, Genome Canada, Health Canada, the Canadian Intellectual Property Office, Environment Canada, the National Research Council, the Canadian Institutes of Health Research, the Canada Foundation for Innovation, the Social Sciences and Humanities Research Council, the Natural Sciences and Engineering Research Council, the Conference Board of Canada, the Canadian Association for Community Living, the Council of Canadians with Disabilities, the Organization for Economic Cooperation and Development, and the World Intellectual Property Office.

A continuing intellectual and personal set of thanks are owed to col- leagues and staff at our respective academic institutions, the School of Public Policy and Administration at Carleton University, the Politics Department, University of Exeter in the UK, and the Faculty of Human and Social Development, University of Victoria.

G. Bruce Doern and Michael J. Prince
January 2012

Abbreviations

ADM	assistant deputy minister
AHR Act	Assisted Human Reproduction Act
AHRC	Assisted Human Reproduction Canada
BGTD	Biologics and Genetic Therapies Directorate
BRI	Biotechnology Research Institute (of NRC)
CADTH	Canadian Agency for Drugs and Technologies in Health
CEAA	Canadian Environmental Assessment Agency
CEDAC	Canadian Expert Drug Advisory Committee
CBAC	Canadian Biotechnology Advisory Committee
CBS	Canadian Biotechnology Strategy
CBSEC	Canadian Biotechnology Secretariat
CCGF	Canadian Coalition for Genetic Fairness
CDR	Common Drug Review
CFI	Canada Foundation for Innovation
CFIA	Canadian Food Inspection Agency
CIHR	Canadian Institutes of Health Research
CIPO	Canadian Intellectual Property Office
DMF	Decision-Making Framework
DNA	deoxyribonucleic acid
EACSR	External Advisory Committee on Smart Regulation
EFSA	European Food Safety Authority
FAO	Food and Health Organization
FDA	Food and Drug Administration
GE³LS	ethical, environmental, economic, legal, and social issues
GINA	Genetic Information Non-Discrimination Act (US)
GM	genetically modified

GMO	Genetically Modified Organism
HPFB	Health Products and Food Branch (of Health Canada)
IDCWG	International Development and Cooperation Working Group
IG	Institute of Genetics
IP	intellectual property
LMO	living modified organism
MRC	Medical Research Council
NABST	National Advisory Board on Science and Technology
NAFTA	North American Free Trade Agreement
NBAC	National Biotechnology Advisory Committee
NCE	Networks of Centres of Excellence
NDP	New Democratic Party
NGO	non-governmental organization
NOC	Notice of Compliance
NPG	new public governance
NPM	new public management
NRC	National Research Council Canada
NRC (U.S.)	National Research Council
NSERC	Natural Science and Engineering Research Council
NSI	national systems of innovation
OECD	Organisation for Economic Co-operation and Development
OTA	Office of Technology Assessment (US)
PBI	Plant Biotechnology Institute (of the NRC)
PTNs	plants with novel traits
R&D	research and development
rbST	recombinant bovine somatotropin
RIAS	Regulatory Impact Assessment System
RSA	related science activities
S&T	science and technology
SSHRC	Social Sciences and Humanities Research Council
WHO	World Health Organization
WIPO	World Intellectual Property Organization
WTO	World Trade Organization
WTO-TRIPS	World Trade Organization Trade-Related Intellectual Property System

List of Tables

THREE BIO-REALMS

Biotechnology and the Governance of Food,
Health, and Life in Canada

Introduction: Three Realms of Biotechnology

Biotechnology 'involves the use of living organisms, or parts of living organisms, to provide new methods of production and the making of new products' (Health Canada 2010a, 1). No longer a new technology, it is still very much a transformative and enabling technology that cuts across many sectors of the economy and society. It functions as an industry and set of products and processes in such areas as food, plants, human health, and the life sciences linked to the human genome. The OECD's sole definition describes biotechnology as 'the application of science and technology to living organisms, as well as parts, products and models thereof, to alter living and non-living materials for the production of knowledge, goods and services' (OECD 2010, 7).[1]

Our conception of biotechnology differs from most other studies in this burgeoning literature. Many studies focus on explaining the fundamental science itself and examining certain historical and contemporary applications; some studies address ethical aspects, while others explore ecological implications. We understand the 'reality' of biotechnology as a constructed and fluid scientific phenomenon as well as an emergent and dynamic field of public policy. To a degree, there is an apparent concreteness, an objective realism to biotechnology: the application of 'hard sciences' in laboratories and other institutions, through research and development, resulting in specific processes and products, with actual effects on humans and natural environments. Yet these very same hard sciences yield unpredictable findings, changeable premises, and, at times, paradigm shifts in our basic understandings of the cosmos and natural world. At the same time, science and the biotechnological do not operate outside of, or prior to the social and the political (Blank and Hines 2001).

Questions of biotechnology are questions of power. Science and biotechnology take place within organizational structures, historical and geographical contexts, and myriad cultural practices, all infused with relations of authority and influence. With their mandates, hierarchical structures, operating rules and procedures, and outputs of products and services, organizations in the economy and society – especially large, enduring, and strategic organizations – are realities external to individuals and groups. Even so, corporate and community organizations are human creations depending upon and deploying both material and symbolic resources. As such, these organizational realities cannot be taken for granted.

A mixture of values, actors, interests, relationships, and circumstances specifically forms biotechnology. Over time, biotechnology governance has congealed into formal structures, recognized norms, and institutional discourses, along with dominant and alternative bases of knowledge. While the form of biotech governance has expanded and solidified, it is not necessarily around an easy consensus and it is certainly not settled. Policy and governing arrangements remain open to competing interpretations and shifting expectations and, therefore, decision-making trade-offs and political paradoxes. The reality of biotechnology is challengeable and thus changeable. As a consequence, there is no single system or essential structure of bio-governance. There are at least three realms of biotech, which we examine in this book, operating at multiple levels of governance and interacting with several forms of power and democratic politics.

Another aspect of the ontology of this book is our conception of the self and the meaning of human life. We view the individual as an interdependent actor, supported and constrained by affiliations and interactions, all located in a given historical period and specific social context. The formation of the self also has a biological materiality that involves having a body along with a sociological reality of being and becoming a body (Berger and Luckmann 1967). In the words of Charles Taylor, 'We are embodied agents, living in dialogical conditions, inhabiting time in a specifically human way, that is, making sense of our lives as a story that connects the past from which we have come to our future projects' (1991, 105–6). The self has a plurality of roles and identities through the life course: family member, student, citizen, worker, consumer, producer, donor, researcher, activist, scientist, and so forth. Each role and identity is constituted through intimate relations and public dealings with others, shaped by cultural, economic, ethnic, gender, moral, political, racial, and scientific features.

The dynamic of support for, and opposition to, biotechnology comes from contending economic and industrial interests, divergent expert scientific interpretations of research studies, and diverse social groups and individuals. As one biotech commentator has succinctly put it, 'The step from "wow" to "woah" is a very short one these days' (Harris 2010, 13). The 'wow' looks upon biotechnology as a marvel of our present era, while the 'woah' likely regards biotechnology as a malaise of modernity. Sutton expresses a similar sense of caution when he says that 'it is hard to deny that with every biotechnical breakthrough, with every potential cure for disease or affliction, there is an accompanying sense of unease' (2009, 2). Biotechnological developments are interpreted as enabling and disruptive, even unsettling or disrespectful in certain contexts.

For many decades, biotechnology has been changing societies, their institutions, and how people think about many economic and social issues. It has generated several ongoing debates about democratic and human choice in Canada and internationally. With respect to bio-food, debate has focused on the need for transparent consumer information and choice about bio-food products and processes that many see as unnatural and that consumers and citizens should not have to consume if they chose not to (Wiles 2007). Other producer interests view bio-foods as being simply novel foods that reflect scientific progress and that confer other economic and environmental benefits.

The gradual but then high-volume emergence of bio-health research and products from the early 1990s on generated debates ranging again from the proclaimed benefits of such products to the ethics of the research that underpinned their development. The politics of embryonic stem-cell research has led to intense partisan and religious debate in the United States, including court battles. Thus far such high-profile politics has occurred to a much greater extent in the United States than in Canada, but this may change in scope and intensity because U.S. developments will affect Canada as actual products emerge (Boyer 2010; Knowles 2010; Kolata 2010). In addition, because bio-health products and processes involve extensive increased patenting and intellectual property rights, opposition centres often on the commodification of health and life and how to put limits on it (de Beer and Guaranga 2011).

In the last decade in particular, debates surrounding bio-life products and processes, including assisted human reproduction, genetic testing centred even more on the nature of life itself, in what Fukuyama (2002) has labelled our 'post-human future.' This focus of debate sharpened especially after the mapping of the human genome early in the twenty-first century, and the resultant changes in research, diagnostic tools, and approaches to manage and/or cure diseases and in the ways in which it may allow humanity to reproduce human life itself (Caulfield, Ries, Ray, Schuman, and Wilson 2010; Collins 2010; Hochedlinger 2010; Shreeve 2004).

Over the last three decades, biotechnology as a whole has also involved debates about the nature of innovation and the vision of an innovation economy and society. Debate about biotechnology, increasingly linked with other transformative technologies such as information technology and nanotechnology, centred on those who argued for the need to foster biotechnology firms, research, products, and processes so that Canada could compete in a global knowledge-based economy. Restraining opposition voices focused on the need for better and more democratic technology assessment and for greater care about what innovation actually is and which interests benefit from it or are actually or potentially harmed by it.

These and other major debates about biotechnology are engaged in this book's analysis, including what the debates and their outcomes reveal about contending ideas, values, power, and democracy in the Canadian biotechnology story.

Purposes and Contributions

In *Three Bio-Realms*, we are crucially interested in building on and contributing to these key debates. Our focus is more precisely and mainly on the governance of biotechnology in Canada set in an international context. *Governance* refers to the separate and joint actions taken by the territorial state (or states) and by interests, networks, institutions, and individuals, in the private and social spheres, that interact with the state and whose cooperation and coordinated actions are needed to succeed in modern policymaking and implementation (Aucoin 2008; Bellamy and Palumbo 2010; Flinders 2008; Peters 1999; Rhodes 1997). Governance, therefore, is broader than government and broader than public policy by itself. It can be variously and at different times be state-centred or emerge gradually from outside the state.

More particularly, the book has three purposes:

- To critically map and explore the complex functioning of the Canadian biotechnology governance regime over the past thirty years in the emergence of three shifting, reinforcing, and colliding governance realms of the bio-economy and bio-society: the bio-food realm; the bio-health realm; and the bio-life realm. For the regime overall and for each of the three realms we map and discuss the key ideas, systems of power and democracy, agency structures, processes, policy instruments, and technologies at play.

- To understand the governance of this transformative technology in the underlying dynamics of interests in network-based democracy that seek both to support biotechnology and to regulate it in the public interest both inside the state and at the broader co-governance levels of business interests, research institutions, non-governmental organizations (NGOs), and personal choice and identity by individuals as consumers and citizens. Regarding interests and networked-based democracy, we trace the changing nature and complexity of the key interests tied to an appreciation of networks as a societal mode of organization, as policy-mandated requirements, and as a new form of complex accountability and bureaucracy. With respect to democracy, we probe some of the claims about the adequacy of democracy in the biotech governance regime when democracy can mean quite different things, including representative Cabinet-parliamentary democracy, federalist democracy, interest group pluralism, civil society democracy, and direct democracy involving citizens as active subjects engaged in self-regulation, in focus groups and social networks, and as environmentally and health-conscious consumers.

- To place the Canadian biotechnology governance regime in the context of national and international influences and power informed by changing ideas about science-based governance, precaution, and risk-benefit regulation in both pre-market and post-market phases of production and use and also centred on research support and funding in the name of an innovation-centred economy. This involves structures, processes, and ideas that are forged in complex scientific fields, academic disciplines, and public agencies (national and international), and with particular meanings about how to handle new technologies and products in comparison with well-established ones, and to supervise their research ethics and other ethical dimensions.

The book is written for a broad audience of students, practitioners, players, and citizens interested in, and affected by, the bio-economy and bio-society and how biotechnology is governed in the context of highly networked private and public interests.

Building upon other important published sources (academic, governmental, and NGO-related) and on an understanding of the international governance of biotechnology, we saw the need to provide in one book a systematic view of the governance of biotechnology in Canada. The notion of an integrated account is not easy, given that the essence of biotechnology is that it is a fast-moving field of scientific, technological, economic, and human development. Hence there are potentially many different conceptual focal points and themes that one could choose to examine, even within the notions of a governance focus.[2]

Our analysis seeks to answer three central questions: What is the nature of biotechnology governance change? What factors mainly explain such change? And what do such changes tell us about the changing configurations and shape of public and private power as revealed by the Canadian biotech governance regime, as they play out in a set of complex arenas and kinds of Canadian democracy?

The main empirical contribution of the book is that it provides the first integrated mapping and analysis of the political and historical evolution of the three bio-realms over thirty years and hence of the overall Canadian biotech governance regime. The main conceptual contribution is that it develops and employs an analytical framework that systemically explores three crucial but linked elements of biotech governance: the state as supporter and regulator; business, NGO, and research interests and network-based democracy; and science-based governance and precautionary governance.

A related contribution is that this framework enables us to show that three forms of bio-power are present within and across the three bio-food, bio-health, and bio-life realms: business-dominated pluralistic power, networked power, and self-disciplined power as defined below. These in turn function within a larger set of arenas and criteria of Canadian democracy and politics, including representative parliamentary-Cabinet democracy, federalist democracy, interest group pluralist democracy, civil society democracy, and direct democracy (Bickerton and Gagnon 2009; Campbell, Pal, and Howlett 2004; Williams 2009).

Power is a concept central to understanding what biotech governance involves, what bio-politics is, how biotech policy is made and implemented or resisted. Our approach lays emphasis on the multidirectional and relational features of power. There is no single locus of authority or capacity in the biotechnology governance regime. Resources and relations of power that work through the three bio-realms are varied, networked, and resisted, and they change with different levels of capacity for business, consumers, NGOs, men and women, specific ministries, and agencies of government.

What nominally creates the bio-economy and bio-society, and seemingly connects them, are basic and applied forms of scientific research and development with living organisms. At the same time, we underline the effects of the technologies and relations of power and knowledge on bio-policymaking and regulatory governance, whether nationally, continentally, or further afield internationally.

The importance of investigating the exercise and relations of power in examining biotechnology is apparent in intellectual property rights and the patenting of genetic products; in debates over the meanings of personhood, gender, family, and citizenship; in choices of governing instruments to support and prohibit certain activities or products; in the creation of mandates, the exercise of constitutional powers, and the shaping of organizational structures; and in the multiple forms of democratic politics at work that interact to shape some configuration of trust and confidence along with misgivings and cynicism about biotechnology.

Governing instruments in this policy field range from the forceful to the flexible. We will show how power formally anchors state organizations and societal institutions, and how individuals internalize power within belief systems and self-controls. The effects of regulatory power in biotech policy are commonly restrictive *and* permissive, with prohibitive and productive consequences for both public and private actions.

Power relations found within the biotechnology regime are linked to particular forms of knowledge (expert and lay), discourses or political rationalities (official or dominant and oppositional), and social structures (civic, economic, familial, gender/sex, and religious). These biotechnology–power–knowledge relationships play out at both the individual level of the self where people are subject to specific opportunities or risks, products, and rules; and, at the collective level, where people are subjects of funding policies, patent laws, and the general governance of science and technology.

Power resources and effects are rarely distributed evenly. Different interests have distinctive capacities to act, and they articulate diverse ideas and contrasting views of what priorities should be in biotechnology as it applies to food, health, or human life. Each of these views on science and risk, for example – be it the precautionary principle or science-based governance – produces particular power effects. Consequently, there are patterns of power *over* (domination), power *with* (collaboration or partnerships), and power *to* (the capability to do or influence something). At times, the structures and processes of power relations may be experienced as non-zero-sum, whereby all participants feel their voice is respectfully heard and their interests have been substantively advanced. More frequently, however, relations of power come across as zero-sum relationships of starkly forced trade-offs and significantly unequal policy outcomes.

Biotechnology, as well as being a technology of life sciences, is a technology of power relations. For approximately two hundred years, the modern state has been concerned with the administration of public health and regulation of reproductive functions of the population, in short with bio-politics and bio-power or what Foucault (2008) called the political or regulatory technology of life. Thus, the policy and governance developments of the past thirty years that we chronicle overlap with this much longer history of the regulation of agriculture and food, medicine and health, and sexuality, human life, and the population (Blank and Hines 2001).

Our analysis of these three bio-realms indicates that the biological has increasingly come under state control in the late twentieth and early twenty-first centuries. New objects of scientific research and knowledge have been generated, and additional practices and products have become subject to medical, commercial, personal, and governmental controls. Bio-politics concerns the exercise of power and knowledge in producing plant traits, improving and managing personal health, cloning cells, creating human life, and building families. Bio-power involves the authority to make life forms, improve life styles, and regulate life choices. It reflects and shapes the biological materiality of the self as both having a body and being a body. Bio-political issues, such as fertility and infertility, are simultaneously general biological phenomena and matters of personal capacities or incapacities. These issues and power dynamics work at the level of the individual body and family as well as the body politic of society and the economy.

On government and on biotech governance functioning in a set of even larger arenas, we emphasize the exercise of power and broader claims of democracy by interests in both public and private structures and through intergovernmental relations. Government, in the sense of the executive organization and administrative structure functioning in a parliamentary system, is but one part of the state. The Canadian state, and hence its democratic underpinnings, is an amalgam of institutions, some formally constitutionalized and others less formally entrenched but nonetheless ever-present. In this way, we locate bio-power in an overall democratic setting, one that includes the roles of Cabinet-parliamentary government, federalism, the configuration of inter-est groups and interests, civil society groups, and direct democracy citizen–state relationships.

Bio-power, and its sources and methods of working, resides not only in the formal state structures, but also in economic markets, medical-scientific structures, and cultural and societal institutions. The roles of international agencies and treaties, the mass media and the courts are also touched on in our analysis. Thus, beside and beyond the govern-ment of biotechnology is the governance of biotechnology, with multiple forms of power traversing the realms of food, health, and human life.

The Three Bio-Realms and Related Products and Processes

The three bio-realms of bio-food, bio-health, and bio-life are clustered around emerging sets of products and processes. Biotechnology prod-ucts in food in particular have been on the market in some cases for forty years and longer, and areas such as plant breeding have been going on for centuries. The *bio-food realm* therefore includes food crops such as corn, soybeans, rice, and many others. In the *bio-health realm*, products include recombinant blood products, agents used in gene therapies, embryonic stem-cell research, tissue-engineered products, and insulin, to name only a few. In the *bio-life realm*, biotechnology builds on the mapping of the human genome and includes the modern worlds of DNA testing and identification in everything from police enforcement to genome-centred forms of personalized medicine and targeted drugs for smaller sub-populations. It also builds on and includes aspects of old and new technologies for assisted human reproduction.

Research and product development is also shifting from local or regional markets to trade at continental and worldwide scales. As Thacker observes, 'Bioscience research and the biotech industry are increasingly organized on a global level, bringing together novel, hybrid artefacts (such as genome data bases and DNA chips) with new means of distribution and exchange' (Thacker 2006, xv).

While the listing of biotech products and processes seems simple enough in some respects, in other senses the notions of *product* and *process* are varied and have diverse meanings and implications. In the bio-food realm, *product* can mean a particular food one consumes or it can also mean a plant with novel traits released in an unconfined way into the environment. In the bio-health realm, the idea of what a product is can move beyond older conventional distinctions between drugs and medical devices to those that are increasingly a hybrid combined product or involve some kind of research process and genetic testing technique. In the bio-life realm, *product* takes on meanings that lose many of their traditional moorings. While both bio-food and bio-health products can be assessed at the pre-market and post-market phases of assessment and regulation, and even withdrawn from the market or the health and the food system, bio-life aspects cannot be recalled in anything like the same sense. The defective 'product' may be a child or offspring for whom the pre-life or early stages are crucial and need to be effective in profoundly human terms.

The story of these realms is chronological, but not entirely. Different core sets of governance institutions were constructed and imperfectly negotiated as each realm came on stream: first bio-food, then bio-health, and finally the bio-life realm. There are some overlaps and collisions, as the three realms were built as add-on features to form the complex overall biotechnology governance regime that now exists. Thus the regime is not so much an elegantly built and designed one but rather one that at times is the imperfect 'resultant' of complex processes, decisions and non-decisions, rival pressures, and tangled values.

These then are the three bio-realms whose governance and related political features, structures of power, and challenges are the focus of this book. The human genome has been mapped, but the analogous basic mapping of the biotechnology governance regime has not yet been adequately done. The book's biotechnology governance mapping and analysis coverage is purposely and necessarily extensive. At the same time, however, it cannot cover the full range of what one author calls the substrata of biotechnology (Smith 2009). These include environmental biotechnology, biofuels, enzyme technology, and forest biotechnology, taking on complex ranges and forms of bio products and processes. For example, environmentally, bioremediation approaches are used to clean up environmental contaminants and for biomass conversion. Bio-fuels are being increasingly researched and adopted separately or blended with petroleum fuels, and they raise enormous issues about sustainable food production, bioethics, food price increases, and related concerns about agricultural subsidies (Forge 2007; Mitchell 2008; Phillips 2010).

Each of the three bio-realms has different levels of familiarity and clarity. The bio-food realm seems, with hindsight, to be relatively clear-cut or simpler, partly because it was established first and is more solidly a state-centred form of governance. Bio-health is basically the second biotech governance realm to form, followed by the bio-life realm, with overlaps more prevalent between the latter two realms. The bio-life realm, while gestating for some time, is the most dispersed and indefinite in governance and therefore more a work in progress than the other realms, although each has ongoing scientific developments and policy concerns. Issues of life have also been a part of the bio-food realm, at least in the sense that there are concerns about what products are natural, and that plant and animal life are all a part of a living and interdependent ecosystem often referred to in general as the collective bio-commons.

It is not difficult to find everyday examples of these complex and deeply felt issues and debates across each of the three bio-realms in political actions, media coverage, and in the formation of groups and networks. For instance, in the bio-food realm in 2010, an NDP MP's private member's bill passed second reading in the House of Commons but then was eventually defeated in 2011. Bill C-474 sought to protect farmers by calling for an analysis of potential harm to export markets prior to approving new genetically engineered seeds (New Democratic Party 2010a).

In 2009 in the bio-health realm, the Canadian Coalition for Genetic Fairness (CCGF) formed as a collection of fifteen disease-based societies and foundations. Among their goals are to 'explore policy options aimed at mitigating/preventing/prohibiting the use of genetic information in assessments and decisions related to employment and insurance (life, disability, critical illness, mortgage, extended-health)' (Canadian Coalition for Genetic Fairness 2010). In 2010, a private member's bill presented by an NDP member of Parliament, with the support of the CCGF, would amend the Canada Human Rights Act to prevent the discrimination of people based on genetic characteristics (New Democratic Party 2010b).

This phenomenon, studied recently in the Canadian context, found 40 per cent of respondents reporting experiences of discrimination based on family history or genetic test results (Bombard et al. 2009). It is noteworthy that in the United States, more than forty states have genetic discrimination laws that deal with health insurance (and more than thirty also deal with employment). In 2008, President Bush signed the Genetic Information Nondiscrimination Act, to prohibit genetic discrimination by employers, insurers, and unions. In Europe, too, several countries prohibit the use of genetic testing in insurance, among other activities. Canada lacks such laws (Lemmens, Pullman, and Rodal 2010).

In the bio-life realm, there is definitely positive and enthusiastic coverage of the potential health benefits of genomics (Ries and Einsiedel 2010). Media stories trumpet the first sequencing of an entire family's genomes, which researchers regard as leading to a powerful tool for tracking down the defective genes that cause inherited diseases (Sample 2010). But there is also coverage of how consumers, especially in the United States, are also slow to embrace the age of genomics in terms of purchasing diagnostic genetic-testing services, partly as the result of high prices and dubious claims about the value of such services (Pollack 2010a, 2010b).

In May 2010, massive media coverage was given to the announcement by Craig Venter, a key private-sector scientist and entrepreneur in the initial mapping of the human genome, that he and his team of researchers had developed the first synthetic living cell. The researchers copied an existing bacterial genome, sequenced its genetic code, and then used computer-based 'synthesis machines' to chemically construct a copy (Wade 2010b). While this kind of synthetic biology has in some respects been underway for some time (Kuzma and Tanji 2010), the Venter breakthrough is cast by some observers as 'baby steps to new life-forms' (Judson 2010a).

In earlier periods across the last thirty years, other bio-controversies on bio-food, bio-health, and bio-life have emerged and garnered varied media and political coverage. Many of these emerge in more detail in the chapters that follow.

Main Arguments

In this book we advance six main arguments regarding the nature and evolution of the Canadian biotechnology governance regime. Each is previewed here in turn and is developed and empirically examined as the analysis proceeds.

Our **first and overall central argument** is that biotechnology in its main emerging forms is *altering the boundaries between the public and the private, between the economic and the social, and between the scientifically possible and the ethically preferred.* Since the 1980s, although with longer antecedents, three distinct realms of biotechnology have emerged that are generating a shifting, at times reinforcing, and often colliding set of ideas, interests, and relations of power and knowledge. Biotechnology governance is shaping, and being shaped by, relations between and among the state, science and technology, the economy, and society, as well as by multiple ideas and forms of democracy, some classically old and others fairly new, as we struggle to determine the qualities of human life and the meaning of society in the twenty-first century.

Our **second argument is** that *three forms of power are present within and across the three bio-food, bio-health, and bio-life realms: business-dominated pluralistic power, networked power, and self-disciplined power.* They are present in all three bio-realms, yet in widely varying degrees. Moreover, it is not just politics and interest pressures that yield policies or cause policies to emerge. In important ways the intrinsic nature of policy – in this case, biotech policy – helps produce different kinds of politics and therefore different kinds of technology-driven power relations and governance arrangements.

The *business-dominated pluralistic form of power* refers to state structures and a narrower pluralistic set of interest group relations dealing with both regulation and support. It is pluralistic in many of its formal democratic features and aspirations, but in practice it functions in an extraordinarily institutionalized way so as to favour some interests over others in one or more areas of policy and public affairs. This characteristic emerges most clearly in the bio-food realm, where agricultural and business and research support interests are shown to be fairly concentrated and cohesive, particularly in comparison with the mainly dispersed consumer and environmental interests. It also emerges in bio-health and bio-life in the role of interests such as organized medicine and related autonomous health professions, the drug industry, and the health industry where complex forms of medical power are at play. For example, genetic counselling as a practice domain has evolved since the 1970s and includes elements of bioethics, nursing, medicine, family therapy, paediatrics, prenatal genetics, and medical psychology.

Business-dominated pluralistic power in biotech governance also intersects with the institutionalization of gender-structured relations in federal laws and regulations, social beliefs and family practices, and the practices of official politics and grassroots politics.

The *networked form of political power* refers to state–society/economy relations, co-regulation and self-promotion by certain private and research interests. The bio-health realm seems to most exhibit this mode of power. Even if drug and smaller biotech companies are a central interest, literally dozens of networks of science, scientists, and medical practitioners are involved, encouraged, and mandated by policy to be ever more network-like and partnership-based in the way they work and are funded. As well, networks are technically inherent in the actual nature of the scientific-genome structure of genes and links among genes and with diverse complex environments.

The *self-disciplined form of power* refers to governance of the self by the self, influenced to be sure by actions of the state and other actors, yet in significant ways governed by highly personalized forms of power beyond what the state can actually control or have power over. Crucially, this includes gender power, where women in particular seek to empower themselves vis-à-vis men with regard to numerous aspects of reproduction and also the caring of families and friends. We see this especially in the analysis of the bio-life realm and it is also a feature of the bio-health realm. The self-disciplined form of power relates to civic regulation (Prince 1999) and also self-regulation, defined partly to be regulation of the self by the self. This involves the private behaviour of consumers and, more generally, individuals, particularly 'women' as a category, which we deconstruct as persons in certain social roles as 'embodied actors,' such as infertile partner, expectant mother, aging parent or grandparent, and persons with a developmental disability or other impairment, among others. Institutions in which this third form of power operates include families and related kinship relationships, women's organizations and support groups, health clinics, physicians' offices, and hospitals, often with the body as an object of power and related knowledge and research.

The **third argument** we make is that *fundamental political realities explain why the Canadian biotechnology governance regime does not have a central point of governance within the state and why it has rarely been a high political priority.* This seems paradoxical, given the evident transformational nature of biotechnology.

One part of the paradox is that biotech governance has become more complex and even impenetrable over its thirty-year history and that it still has no obvious single institutional centre. From the outset, bio-governance has been constructed in its various realms as a multi-agency appendage and set of subunits to existing federal departments and agencies with earlier, much larger, non-biotechnology mandates. In part, this was initially because biotech industry interests did not want the new technology to be cast as new or threatening, but rather simply as novel adaptations of existing technology, such as novel foods. Later manifestations of agency complexity built on this initial design but then became more complex as new products and processes emerged.

A second part of the paradox is that biotechnology has not reached the pinnacle of the federal agenda. It has scarcely been mentioned or given much profile in Throne Speeches or Budget Speeches since 1980, the central agenda-setting occasions for prime ministers and ministers of finance.

In Canada, biotechnology functions in the subdued middle-worlds of political life and governance for a number of linked reasons and features. These are contrasted in part with a higher-profile political presence at times in Europe, where bio-food opposition was strong and sustained, partly because of the relative absence of bio-food producers compared with those in North America. There has also been a higher periodic presence in the United States, in part because of the stronger role of religion in bio-life issues. The posited reasons and features also include the argument that ministers and politicians may not quite know how to deal with it in a raw political sense. The bio-world involves complex kinds of science and technology that are beyond the capacity of non-scientist politicians to discuss and communicate comfortably and with some clarity. Because bio-technologies and their governance deal with boundary-breaking notions regarding the public and private and the social and economic, they yield further discomfort for politicians who have to try to explain complex forces and values in an otherwise sound-bite age of political discourse and criticism. We comment later on whose interests are served or advantaged by this overall subdued profile on the Canadian national policy agenda and on why a comparative reference to Europe and the United States must serve as a cautionary limitation to these proffered reasons.

Our **fourth argument** is that the state's need to support and to regulate transformative technologies such as biotechnology *has resulted in ever broader notions of what support and regulation involves and also the discourse regarding how to speak publicly about these partly conjoined, partly separate, and often conflicting actions.* All industry in Canada is simultaneously supported by some agency of the state and regulated by some part of the state. Thus biotechnology is not alone in having these core features of governance forged, debated, and changed. Across all the bio-realms and the thirty-year period we examine, there can be little doubt that ever broader notions of what constitutes support and regulation have developed. This broadening has occurred in basic political discourse and agency mandates. It has also emerged in the use of basic policy instruments such as spending, taxation, regulation, and persuasion and mixtures of them. These changes reflect both direct biotech policy influences and changes propelled by several other policy fields that indirectly but still have significant impact on the bio-realms.

The **fifth argument** advanced is that *the increased networked nature of the biotech governance regime's public–private interests and individual citizen identities makes it ever more difficult to judge the various claims about whether the regime overall is democratic, and if so, regarding what kinds of democratic principles and what kinds of outcomes and accountability.* Networked biotech governance, as a descriptor of actual working practices and relations of power and knowledge, can make some claims to greater and higher democracy and inclusiveness. It can also produce impenetrable transaction costs and accountability mazes such that, at times, it produces a complex stalemate and obtuse democracy. Biotech governance in Canada plays out in a larger set of arenas and related values about democracy in the social world. Casting the conceptual net relatively wide, we discuss democracy in relation to five such arenas: representative democracy (in Canada's case, elected Cabinet-parliamentary government); federalized democracy; interest group pluralism; civil society democracy centred on broader non-governmental organizations (NGOs), charities, and religious groups; and direct democracy involving individual citizens engaged in self-regulation, in focus groups and social networks, and as environmentally and health-conscious consumers.

Bio-governance includes all of these to some extent across all three bio-realms, yet they emerge with greater salience in some more than others. The fivefold nature of arenas and norms of democracy itself makes assessments of bio-democracy difficult, often because advocates of one arena are silent about the claims and legitimacy of the others.

Our **sixth and final argument** is that *science-based governance norms still underpin the biotech governance regime more than do the norms of precautionary governance. These norms, often seen as polar opposites, are not universally used, but, as we will show in some depth, these two different approaches do exist and also converge and collide.* The notion of science-based governance has itself taken on extended meanings and needs, including evidence-based and intellectual property–centred assessments of what constitutes legitimate 'invention.' It applies differently to law and regulation-making on the one hand, and product assessment on the other. In product assessment, it applies to multiple pre-market regulatory product assessment and approval cycles, and full product life cycles. It applies also to the overt use of research funding to support the technology and to the use of increasingly complex post-market phases of biotechnology – hence the need for other expanded evidence and knowledge-based forms of review.

Precautionary governance norms certainly are an important part of the analytical story. The precautionary principle emerges from environmental ideas, laws, and policies, both international and Canadian. However, our governance-focused analysis shows that precaution as an idea and a human instinct may be induced and produced not just by the statute-based precautionary principle as such but even more so, albeit imperfectly, by the sheer complexity of the biotechnology governance regime. A central feature of this complexity arises from the folding in of other key values centred on ethics in biotechnology research, other democratic arenas of real or partial technology assessment, and the use of biotechnology in particular products and services.

Foundational Concepts and Analytical Framework

The book draws on three foundational concepts and their related strands of literature and analysis: governance and regulation, interests and networked democracy, and the nature of biotechnology and science-based policy. We discuss these literature strands in chapter 1, and they also underpin the historical account of bio-policy in chapter 2 and, of course, the book as a whole. As with all conceptual literatures, the strands have boundary limits but are nonetheless important foundations for the analytical framework we develop and employ to understand the Canadian biotechnology governance regime politically and practically.

That analytical framework, set out in detail in chapter 1, consists of three elements: the state as supporter and regulator; interests and networked bio-governance and democracy; and science-based and precautionary governance.

The *state as supporter and regulator* is a crucial element to understand. Biotechnology is not the only policy and governance field where the state is engaged in, and conflicted by, its need both to support such industries and research and to regulate them in the public interest. Different departments and agencies of the government take on these supportive and regulatory roles, and the interplay and relative influence and power among them must be examined.

The second element in our analytical framework, *interests and networked bio-governance and democracy*, zeroes in on the changing structure of the diverse interests as they function via increasingly complex and fast-forming networks of biotech governance and democracy. The need to see interests as increasingly wanting to, and having to, function in networked contexts and arenas is crucial to understanding biotech governance, various claims about the presence or absence of democracy and accountability, and science-based governance as well.

Science-based and precautionary governance is the third analytical element and refers both to these different ideas and to their related institutional processes. Science-based governance tends to become an increasingly significant norm in international trade law and it also underpins core notions of how to regulate biotechnology products (and of course products in other policy fields as well). For example, we will show how notions of evidence-based governance are being added to the larger science-based concept, as are assessment processes for patents and related concerns about the meaning and limits of patents (Castle 2009; Mills 2010). Precautionary governance is a norm and process that emerges more from international and national environmental policy and law, where the operative instinct is to advocate a preference for the precautionary principle to prevent action and development from occurring until one has better and more complete systems of science and related knowledge in place regarding risks, benefits, and long-term consequences. This too has entered the three bio-realms. Clearly there are interactions, overlaps, and collisions among these framework elements, to which we draw attention throughout the book.

Epistemologically, as researchers we have engaged with this topic for a number of years, conducting interviews with officials in governmental and scientific settings; producing reports on biotechnology for advisory groups; working with social groups and NGOs; and making presentations to parliamentary committees on these issues. Our aims are descriptive and normative: to understand the development of Canadian biotech policy and governance over the past three decades, placing those developments in the national and international contexts; and to present a critique of practices and policy directions, recommending reforms that will deepen the democratic possibilities of biotech governance. Our epistemological stance for that reason is explicitly value-based, informed by norms about democracy as understood in our time in this country.

Methodologically, we have primarily used an inductive form of logic whereby concepts such as supportive governance, categories such as the three bio-realms, and our main arguments have emerged from our analysis of the information collected. In our qualitative research inquiry, we focus on describing rules, discourses, and processes, and on understanding the relations of power, the meanings of events, and the implications of bio-governance for democratic politics. Our exploration has entailed reviewing Canadian and international government reports and public studies on biotechnology and its governance by varied interests covering the thirty-year period. We draw extensively on academic literature from the key literature strands previewed above. Documentary and literature sources were complemented by over one hundred interviews, which the authors conducted on a not-for attribution basis with officials, scientists, interest group spokespersons, and other experts knowledgeable about biotechnology and its governance regime. Conducted over several years as the three bio-realms emerged and evolved, the interviews provide an important context and subtlety to complement the insights gained from the primary sources and our own experiences as public policy, governance, and regulatory researchers.

Structure of the Book

Three Bio-Realms is organized into two parts. Part One provides the underpinnings and analytical framework and biotechnology policy history. Part Two takes the reader into the three shifting and evolving bio-realms in Canada and also the changing supportive aspects and agencies of biotech governance.

Within Part One, chapter 1 fleshes out the analytical framework previewed above. Drawing further on the three core foundational literature strands, it sets out why the three governance elements are central to understanding the governance of biotechnology in an integrated historical, technological, policy, and political manner.

Chapter 2 provides a brief historical view of the evolution of Canadian biotech policy, including the partisan political preferences of successive federal governments and how it links to international policy and pressures. The history of biotechnology over the past thirty years is not some natural progression or coherent whole built upon ever-expanding scientific research and technological applications. Nor is biotech policy just an undertaking of curious inquiry and rational progress. It is a political phenomenon as much as anything else. Biotechnology derives from several contextual and contested forms of knowledge and discourses as well as relations of power, nationally and internationally.

Within Part Two of the book, chapter 3 provides a portrait of the federal biotechnology governance regime as a 'designed' but also a somewhat haphazardly negotiated, pressure-produced, and changing regime. It covers some of the regulatory aspects, but the focus is more on the supportive research governance of the overall regime as well as the biotech advisory bodies and arenas of overall debate, consultation, and engagement, and imperfect technology assessment processes.

The next three chapters look more closely inside the three shifting realms of Canadian biotechnology governance. Chapter 4 focuses on the bio-food realm, in particular the regulatory processes and dynamics of bio-food product assessments and approvals set in the larger context of novel foods science-based regulation. The bio-food realm evolves in the context of agri-food business pressure and power to promote such products in concert with federal departments such as Agriculture and Agri-Food Canada and federal research establishments and funding. It also evolves amidst counter but weaker pressures from consumer interests and parliamentary committees regarding individual identity concerns about the role of choice in using or not using bio-food products.

Chapter 5 zeroes in on the bio-health realm of the biotechnology governance regime, including its somewhat later emergence as a partial 'add-on' to the initial bio-food realm. The chapter focuses on the broadened set of regulators and supportive elements involved, the complex dynamics of product approvals and funding, as new high volumes and much more tailored products emerged, as well as embryonic stem-cell research, some genome-centred aspects of genetic testing, and personalized medicine. Personalized medicine refers to the ability to classify individuals into sub-populations that differ in their susceptibility to a particular disease or their response to a specific treatment. The networked democratic interest structure and interplay in the bio-health realm is more complex, often because the politics of bio-health is somewhat more positive for politicians and citizens than in the bio-food realm, which formed earlier.

Chapter 6 then looks at the broader bio-life realm of the biotechnology governance regime. In some respects this realm has inevitable crossover points with the bio-food and bio-health regimes. In a real sense it represents, captures, and seeks to find, often in a catchup manner, its own governance and political space in a system that was forged and congealed earlier. An array of interacting and colliding issues emerge in this third realm centred on bio-life. The chapter focuses on the nature of – and delays regarding – the establishment of agencies to regulate reproductive technologies. It also considers genome-related products, processes, and debates involving personalized medical treatments and genetic testing, including related notions of human cloning and further commentary on stem-cell research and ethics.

Chapter 7 looks across and compares the three bio-realms in relation to each of the three elements of the framework: the state as supporter and regulator, interests and networked biotech governance and democracy, and science-based and precautionary governance. It presents our main conclusions on the evolving nature of the governance of biotechnology in Canada set in its international context, on our six main arguments, and on the future governance and democratic challenges to be faced and needing reform.

In Canada and internationally, the biotechnology journey and the resultant emergence of both a bio-economy and bio-society involve a series of governance and democratic issues that range from science support and ethics, to regulation in a range of public and private interest contexts, to intellectual property, innovation, and the commodification of life and health, and to international trade and relations among trading blocs and developed and developing countries.

PART ONE

Analytical Framework and Historical Policy Context

1 The Canadian State, Networked Democracy, and Science-Based Governance

Introduction

Our goal is to comprehend and explain the changing nature and structure of bio-governance, which we characterize as three shifting bio-realms: bio-food, bio-health, and bio-life. The three central questions we examine are: What is the nature of biotech governance change? What factors mainly explain such change? And what do such changes tell us about the changing configurations and shape of public and private power as revealed by the biotech governance regime, as it functions within a larger set of arenas and concepts of democracy in Canada?

To begin answering these questions, this chapter discusses the conceptual foundations that underpin our analytical framework for understanding bio-governance in Canada and sets out the analytical framework in detail. The larger part of this chapter presents our focus on the three main governance elements: the state as biotech supporter versus biotech regulator, networked biotech governance interests and democracy, and science-based and precautionary governance. In each of these governance elements, policies, policy ideas, and biotech content are embedded and disputed by public, business, and social interests, and individual Canadians reacting to, and trying to shape, national and international biotech governance institutions. The underlying conceptual foundations and their related literature strands are discussed in the shorter conceptual context-setting first section of the chapter. Two of these are initially broader than biotech per se and emerge from the study of policy, governance, and politics in many fields. The third conceptual foundation is much more specific to biotechnology and is necessarily coupled as well with much broader aspects of science-based governance.

Conceptual Foundations

The study builds on three conceptual foundations and related literature strands, each of which broadly contributes to the selection and design of the three main elements in the analytical framework and to our overall narrative. Each conceptual foundation and related literature strand is examined for theoretical ideas and empirical reference points. Both comparative and Canadian, the three conceptual foundations encompass governance and regulation, networked democracy and interests, and the nature of biotechnology and science-based governance.

Governance and Regulation

Governance as a concept emerged in the literature on politics, policy, and public administration over the past thirty years (Aucoin 1997, 2008; Rhodes 1997). In one sense, its meaning can be expressed simply as an effort to recognize more explicitly that governance was more than government, more than the state, and more than public policy pronounced and implemented by the state and its bureaucracies. Thus gradations emerge in that governance can be quite state-centred but also, at different stages of development, more arm's length and socially and economically rooted. It also implies that the state plays a role characterized at times more by steering than rowing, and also employs softer instruments of governing rather than harder command-and-control regulatory ones.

At its core, governance is a set of interacting policies, decisions, processes, laws, and values involving joint action by the state and its agencies, business interests, non-governmental organizations (NGOs), and civil society interests and entities. It is also featured by great differences and inequalities in power, influence, and capacity among these agencies and interests, with business interests typically having the greatest staying power among private actors in the long game of governance.

As a form of increased and complex delegated authority, governance is influenced by concepts such as the new public management (NPM), whose essence is to make government less hierarchical and top-down and more sensitive to citizens, customers, and clients by employing alternative service delivery modes of organization, and more complex forms and arenas of accountability, greater transparency, and better performance management (Aucoin 1997, 2008; Rhodes 1997). NPM is also allied with neoliberalism with the latter's calls for smaller government, lighter regulatory burdens, and sharper focus on international market competitiveness and benchmarking.

More recent emphasis has been given to concentrations of power in and around the prime minister vis-à-vis Parliament, the Cabinet, and the bureaucracy. Aucoin (2008) refers to this as a shift from NPM to the 'new public governance' (NPG) centred on exceedingly detailed prime ministerial control of policy, politics, appointments, and political communication. In Canada, this has been especially noted in the Harper era and in the context of minority government, but some of these developments pre-date Harper and are globally widespread (Doern and Stoney 2010; Savoie 2010). These changes are seen as being particularly to blame for a perceived weakening of the democratic and representative role of Parliament and of the Cabinet.

Governance is without doubt increasingly imbued with, and linked to, arrangements in global authority and control. The logic of these globalization-propelled international developments has meant that governance is self-evidently multi-level governance (Bache and Flinders 2004). Multi-level governance as a related conceptual construction emerged partly from the unique example of the European Union as a club of nation states but it also resonated in political federations such as Canada where federalism had always meant that multi-level governance is a constitutional certainty and practical reality, marked by controversy, jurisdictional tensions, key areas of cooperation, and joint action.

Governance is also comprehensive in its use of the main instruments of policy. Governance includes the use of taxation, spending, regulation, and persuasion, and all of the subtypes within each major instrument. In short, it involves the deployment of the full set of tools in the governing toolbox (Doern and Phidd 1992; Howlett and Ramesh 2003; Pal 2006). In recent decades, governance further implies a partial relative shift in the mix of instrument deployment towards softer tools of guidelines, codes, and incentives, and systems of public information, education, and persuasion (Prince 2010; Thaler and Sunstein 2008).

Regulation and ultimately regulatory governance has been conceptualized in similar ways. Regulations are 'rules of behaviour backed up by the sanctions of the state' (Doern et al. 1999, 5). Such state-sanctioned rules are variously expressed through laws, delegated legislation (or the 'regs'), guidelines, codes, and standards, and thus there are levels and many types of rule-making, even in the core definitions of regulation. Guidelines and codes, for example, are often seen as realms of 'soft law' or rule-making in the shadow of the law, and that is therefore several steps removed from parliamentary or executive-level central agency scrutiny.

A focus on regulatory governance recognizes that regulation has to be conceived as rule-making processes, outputs, compliance, and outcomes that can emerge from varied top-down, bottom-up, and negotiated processes within the state, among states, and among economic and social interests and the professions. In short, regulatory governance conjures up diverse notions of regulatory procedures, co-regulation, rule-making spaces, and effects (Baldwin, Scott, and Hood 1998; Doern 2007; Hancher and Moran 1989; Radaelli and DeFrancesco 2006).

Multi-level regulation and regulatory governance is a part of everyday governance (Doern and Johnson 2006). Multiple levels of rule-making and product approvals, and the potential and actual coordination and congestion challenges they create, are seen by many economic interests as an inhibitor of economic growth, efficiency, and innovation in both domestic and global markets. At the same time, such levels of rule-making are avenues to enhance democracy and, through them, to pursue health, safety, environmental- and fairness-related public interest purposes that have different potential impacts in different territories. Multi-level regulation also has the potential to increase or confuse basic aspects of democratic accountability, public trust, procedural transparency, and the institutional legitimacy of public and private actors (Doern and Gattinger 2003; Flinders 2001, 2008; May 2007).

We argue that the central pressure to merge, collapse, or rationalize levels of regulation is driven mainly by business interests and market-based ideas of efficiency and liberalized trade. A contributing pressure stems from related technological changes including Internet and virtual reality governance possibilities, recognizing that these too are mediated through particular configurations of power, interests, and ideas (Borins and Brown 2008; Doern and Johnson 2006).

This drive to streamline the multiple levels of regulatory governance is tied closely to related market concerns in Canada about national economic productivity compared to the United States and European Union, not to mention economic powers such as China and India, and about the costs of state regulation and so-called cumulative regulatory burden. In a post-9/11 era, issues of security and terrorism, and thus border issues, are interwoven with these core business and economic agenda concerns. Taken as a whole, this market-based pressure and discourse manifests itself in the call by commercial interests and other advocates to end the 'tyranny of small differences' in rules and compliance requirements between countries, regions, sub-national governments, and even among cities.

This corporate political pressure is juxtaposed against other demands and values of democratic governance in regulation. The growing voice and pressure of NGOs, women's groups, civil society and coalitions, and consumer interests operating at all levels of regulatory governance and through Internet mobilization are crucial as well. There are inevitable issues about inherent institutional inertia and failures to learn and adapt to new fast-moving political-economic circumstances. This led globally to overall efforts to foster 'better' regulation or 'smart' regulation, where innovation agendas are an explicit force for regulatory change (Doern 2007; Doern and Johnson 2006; Radaelli 2006; Radaelli and DeFrancesco 2006). Concerns about innovation and liability in biotech regulation and assessment criteria are also being raised (Smyth 2010).

In this context, it is useful to stress that innovation agendas are often presented as features of post-industrial capitalism – a knowledge-based economy that is somehow a different form of capitalism – and that biotechnology is a key example of this change. Liagouras (2005) takes note of this argument but makes the case that biotechnology and other enabling technologies have much continuity with earlier industrial capitalism, especially regarding private property rights, the commodification of such rights, and core business-government relations.

Regulatory governance and rule-making emanating from the bureaucracy has been challenged by parliamentarians seeking more transparency and civil society engagement. This has particular relevance to the bio-life realm, where assisted human reproduction regulations were examined by a parliamentary committee before being enacted (Dewing 2008). It also has occurred in the bio-food realm where issues regarding consumer rights were raised often but in Canada without much success.

Regulatory governance additionally implies complex rule-making, product approvals, and enforcement in complex overlapping regimes of horizontal policy. Illustrative examples are trade, intellectual property, and environment, as well as vertical sectors, industries, and product groupings. The regulatory literature draws necessary attention to the different nature and dynamics of regulatory agency reputation and power as well as pre-market regulatory approval processes versus post-market monitoring, and indeed in regulatory reform advocacy that seeks to impose the full regulatory life cycle of products and processes (Carpenter 2010; Doern and Reed 2000; Hood 1998; Pal 2006). We take up these important issues in later chapters centred on the three bio-realms. Regulatory governance literature also highlights the elementary need to factor in the considerable differences in product volumes and the complexity of products that face different regulatory bodies in different policy fields.

Regulatory governance additionally involves approaches and actions expressed as civic regulation, self-regulation, and the regulatory craft (Prince 1999; Sparrow 2000). These concepts suggest that regulation entails a delicate and contentious dynamic among governmental rule-making, self-regulation rights and practices exercised historically by professions such as medicine and law, and forms of self-rule and restraint centred on individual and family preferences and community norms and values. The growing complexity of rule-making has also raised arguments that governments need to adopt more overt and fully and democratically debated regulatory agendas and overall regulatory budgets, as they already do for public spending and taxation by government (Doern 2007).

Networked Democracy and Interests

The second conceptual foundation is lodged in concepts and literature on networked democracy and interests. Networks have received greater analytical attention in the study of economic and social institutions and thus hold considerable importance for understanding contemporary debates on, and reforms to, governance and democracy. In political and policy analysis, networks were initially cast by some authors as a particular inner feature of broader arrays of so-called policy communities (Coleman and Skogstad 1990). Policy communities were developed as an analytical category that went beyond traditional interest or lobby groups. Later analyses have broadened networks to some extent (Howlett 2002; Howlett and Ramesh 2003; Montpetit 2004, 2009; Pal 2006).

Meanwhile, in economic and related fields, networks were being analysed in a much broader context, where they were contrasted with markets and hierarchies as basic modes of social and economic organization (Agranoff 2007; Taylor 2001; Thompson 2004; Thompson et al. 1991). *Hierarchies* are associated with bureaucracy, especially traditional Weberian state bureaucracy; hence, with systems of top-down superior-subordinate political and administrative relations accompanied by formal rules, with related forms of civil service bureaucracies essential to representative Cabinet-parliamentary and other systems of democratic government (Hood 1998). *Markets* are organized on the basis of 'voluntary' means of exchange tied to money, commerce, and the making of profits, but with key rules and protections for property rights and transactions provided by the state. *Networks* are contrasted with both of the above in that they are forms of organization characterized by non-hierarchical and voluntary relations based on trust and commonality of shared interests and values where profit is not a defining characteristic (Agranoff 2007; Thompson 2004).

Some attributes of networks as an institutional mode are expressed in terms of constructing *partnerships*. Such partnerships can be truly voluntary but more often they take on the form of policy-induced or required contractual or quasi-contractual partnerships between/among public and private sector entities and interests. In this context, they acquire some of the characteristics inherent in markets or hierarchies as well (Hubbard and Paquet 2007; Kinder 2010).

The Internet as a defining, increasingly dominant, and enabling technology has become a crucial engine of social and economic network formation and for designed systems of e-commerce and e-governance. For many public and private interests, the Internet has greatly reduced the costs of communication and joint action. It has also fostered new avenues for direct democracy by individual citizens, including via social networks (Borins and Brown 2008). Moreover, in commercial terms, Internet development fosters arguments that the 'wealth of networks' is the new driver of technology and profit and of democratic participation (Benkler 2006).

Analyses of networks in public policy and in the conduct of science and research bring out the tendency for networks to be ever more complex, embedded with transaction costs and related layers of bureaucracy – not necessarily bureaucracy of the hierarchical kind but rather of a more horizontal, transactional, and vertical kind (Doern and Kinder 2007; Doern and Levesque 2002; Flinders 2008). These complex networking and partnership arrangements generate accountability regimes that are intricate and produce the need for multidirectional reporting and answerability systems – in short, accountability 'up' to ministers and parliaments, accountability 'down and out' to clients/partners, and accountability 'across and among' numerous entities, including across different parts of government (Flinders 2001; Hill 1999; Taylor 2001).

Links between networks and democracy necessarily involve a practical appreciation of contending notions of public power, formal interest groups, and interests such as individual firms or government agencies that have the power to take action rather than just influence actions. Claims about democratic governance in any policy field, we submit, have to confront the existence of at least five arenas and kinds of democracy (Doern and Phidd 1992; Dryzek and Dunleavy 2009; Pal 2006). They are set out here for analytical purposes and conceptual clarity. In practice they interconnect in numerous ways.

There are five arenas and related different criteria of democracy:

1. In *representative democracy*, which in the Canadian case is Cabinet-parliamentary democracy, the public interest is claimed to reside in elections and majoritarian decision-making but normal majority governments are replaced by minority or possibly coalition governments, and prime ministers may centralize power also in the name of elected democracy. Representative democracy is intricately tied to political parties and their internal system of democracy, and consequently to the role of partisan considerations in policy formation and opposition.

2. In *federalized democracy*, politics and policy divide constitutionally between national and provincial governments, yielding both conflictual and cooperative joint action and various kinds of bilateral and multilateral federalist bargains and diverse views of democratic action.

3. In *interest group pluralism*, democracy is said to be shown through the continuous interplay of interest groups of numerous kinds involved in lobbying, engagement, and consultation with government and each other. Democracy is said by many to be the simple resultant outcome of this interplay. Others see inequalities of power and cast pluralism not as something that achieves such benign results but rather is reshaped into corporatist forms where business interests in particular are more dominant. In the three bio-realms these interests and the power they wield also include professions such as doctors and related autonomous health professions such as nursing and prenatal genetics, wielding significant medical power, but linked to an expanding realm of wider health-industry power that includes university researchers in health disciplines, and drug and diagnostic companies. It also includes the power and influence of patent lawyers and intellectual property professionals.

4. In *civil society democracy*, democracy is said to arise if broad social NGOs of numerous kinds are involved, including disability, environmental, human rights, indigenous, and women's groups, principally those representing the weak and marginalized in society, many of which take up the language of equality-seekers and rights-holders in general and also under the Charter of Rights and Freedoms.

5. In *direct democracy*, democracy is said to occur when individual citizens – voters and non-voters – express their views and have influence through their own individual actions, and through polling, focus groups, social online networks, as citizens and as consumers functioning in the marketplace as environmentally and health-conscious consumers. They also include individual scientists involved as critics and even as whistle-blowers within government and private firms. These forms and forums of direct democracy are all present in the story of the three bio-realms.

Like all aspects of democracy, direct democracy is evidenced by dispute about its nature and its presence and has a deep historical lineage such as the direct democracy model of Switzerland, the practice of referendums in many U.S. states and to a lesser extent in a few provinces in Canada. In the bio-life realm these kinds of direct democracy can also include choices by individuals where the body is a site for reframing and defending one's self-identity in ways that are similar to those examined by social policy scholars in the politics of disability (Caulfield 2003; Montpetit 2003; Prince 2009). We examine this further crucial kind of direct democracy as a form of regulation of the self by the self.

While *interest groups* (business-based, NGOs, and charitable organizations) receive explicit mention in only one of the above notions of democracy, involvement of interest groups is certainly pervasive and complex in Canadian governance and politics. Moreover, interest groups must also be differentiated from *interests* (Doern and Phidd 1992). Interest groups typically have the ability to lobby and influence, whereas *interests* are players who have actual power to act or not act in concrete circumstances, like individual corporations, which can decide to invest or not invest in Canada or elsewhere. Thus, as already noted above, there is considerable political economy and policy studies literature that points to business interests as having disproportionate power. The notion of interests also includes particular individual government agencies that have statutory authority to act and also have wide ranges of discretion regarding when, where, and how to create and enforce rules or dispense funds.

The above conceptual and practical arenas of increasingly networked democracy yield many shifting public-private-personal network coalitions as the definitions of policy problems, opportunities, crises, and agendas change across time. While each arena of democracy generates and anchors core criteria of democracy, it also generates dispute over whether the criteria are being met or whether other democratic arenas are inherently superior as democratic anchors for advancing specific values and actual causes.

The Nature of Biotechnology and Science-Based Governance

The third steam of literature underpinning our analysis conceptualizes the changing nature of biotechnology itself as a science, a set of technologies, and a complex mix of products and processes with intricate links to diverse kinds of science and science-based governance.

Biotechnology as science and technology and a set of products initially takes root in the period covered in this book as bio-food. As the history of biotech policy in chapter 2 shows in more detail, bio-food is immediately anchored in critical discourse that, among other labels, calls it 'Frankenfoods' but simultaneously, in national and international policy and regulatory circles, casts such products as just another kind of 'novel-food' (Canadian Biotechnology Advisory Committee 2002; Mehta 2009). In addition, the core bio-food literature links biotechnology to positive engineered changes in food such as the elimination or reduction of harmful attributes or, alternatively, and more critically, to products that are in some fashion *unnatural* (Wiles 2007). Such bio-foods are also increasingly compared to *natural* organic foods and food production processes, and, in terms of markets, linked to consumers' freedom to choose their own food on the basis of proper public information and real choice (Hickman 2009; Wiles 2007). Bio-foods also provoke religious concerns about the place of certain foods in ethnic and religious groups' historic and current traditions (Brunk and Coward 2009).

Biotechnology and bio-health product literature clusters somewhat later in the period being covered but, as we have noted previously, particular individual bio-health products and processes have been present for some time (for example, DNA-profiling in police services). The bio-health literature comes from sources and policy realms well beyond bio-food's agriculture and food industry boundaries and emerges to a much greater extent from health experts, professionals, and practitioners (Bauer and Gaskell 2002). This literature deals with bio-health products, many of which are seen as potentially very positive for individuals and health-care systems and very profitable for firms and universities engaged in their development (World Health Organization 2002).

Related literature in bio-health deals much more directly than for bio-food with linked issues of patenting, research ethics, stem-cell research, and paying for high volumes of new niche drugs and products under an always strained health-care budgeting system. Similar although even broader concerns emerged following the full mapping of the human genome in 2003 and the resultant promise of a new genomics-centred era of personalized medicine – specifically legitimate concerns about the need for protection against genetic discrimination in health, employment, and insurance markets (Boyer 2010; Collins 2010; Harmsen, Sladek, and Orr 2006; Personalized Medicine Coalition 2009; Williams-Jones 1999).

A further strand of biotechnology literature (and on related regulatory governance literature) addresses broader environmental health concerns, including bio-science's potential use in environmental bio-remediation technologies and processes, such as its use in cleaning up oil spills and other adverse pollution realms. Bio-fuels have emerged as an analytical focus, both supportive and critical. Environmental aspects also arise more negatively in managing the international transportation of living modified organisms or LMOs (Andrée 2009). Here the environmental concerns about bio-products generate criticism about adverse impacts in global trade when products are not policed and monitored effectively and transparently.

More broadly still, criticism has strongly and convincingly come into view about the extended private ownership of what many argue should remain in human terms as the bio-commons, that is, as an overall public good. Criticism is also focused on the related crucial underpinnings of biodiversity in the environment, and the excessive creation of patentable property rights about the human body and its DNA make-up.

Biotechnology in the bio-life sciences is of course intricately linked to science and reproductive technologies with close links to impacts on and the rights of women (Fitzpatrick 2001; Jones and Salter 2009; Mykitiuk, Nisker, and Bluhm 2007; Royal Commission on New Reproductive Technologies 1993). More recently, biotechnology has become deeply entwined with the mapping of the human genome (Boyer 2010; Judson 2010b; Sulston and Ferry 2002; Sutton 2009; Thacker 2006).

This latter literature both traces the 'genome war' as public science and private science competed to be the first to complete the initial mapping, and it brings out the far larger issues of a possible 'post-human' future based not only on cloning and stem-cell research but also even more crucially in the introduction of much more personalized medicine (Fukuyama 2002; Shreeve 2004). Not surprisingly, this literature evokes concerns that range from religious values and the meaning of life, to deep senses of individual identity about one's own body, especially by women, but also among other citizens as individuals and carers (Hauskeller 2007; Knowles 2010; Morris 2007).

Biotechnology literature also points to arguments that in all forms of DNA, genes, seeds, and plants, biotechnology is inherently networked and thus does not consist of discrete technologically determined cause and effect, or definable products or sequences (Bollier 2002; Dutfield 2008; Peekhaus 2010; Van Tassel 2009). Such arguments weave into debates about public versus private science, and, in the realm of intellectual property and patents, about what constitutes creativity and invention and about the need, as stressed above, to preserve and freely share common public pools of knowledge.

Biotechnology and patenting as a form of 'invention-based governance' turns on differences of view and political power regarding each of the main criteria applied to decisions on the granting of patents, such as novelty, utility, and non-obviousness (Bettig 1996; Carolan 2010; Drahos and Braithwaite 2002; Polk 2003; Thurow 1997). Over time, some studies and reports, especially by governments and international agencies such as the OECD, have come to focus on broader forms of bio-discourse (OECD 2005). Partly to move away from the political negatives of bio-food (such as Frankenfood), certain governments or industry departments within governments began to speak more broadly and positively of the bio-economy and even the bio-society, linking it as well to innovation in an increasingly knowledge-based economy.

The literatures and conceptualizations of science-based governance must be examined in the three bio-realms in particular ways in relation to the types of science, technology, or innovation policy and activity, as well as interest group power and network-based influences that might be involved (Doern and Kinder 2007). For example, science-based governance cast as *science and technology* (S&T) policy refers to general government policies to encourage, support, and manage the development of the national scientific enterprise and the education and training of S&T personnel (that is, policy *for* science). S&T policy also promotes and governs the use of scientific and technical knowledge 'in' public policy and regulation where governments need to draw on their internal S&T or the S&T capacities of others to carry out their responsibilities under laws, rules, and international agreements, especially in public interest areas such environment, health, and safety policy and regulation (that is, science *in* policy) (Doern 1972).

When cast as *research and development or 'R&D,'* science is viewed as 'creative work undertaken on a systematic basis to increase the stock of knowledge including the knowledge of humans, their culture and society, and the use of this knowledge to devise new applications' (Industry Canada 2002, 26). In R&D there is an important element of novelty and thus uncertainty in the results. The federal R&D definition is drawn from the main global reference, which is the OECD Frascati manual, published initially in 1963, but with revisions of the original text.

Crucially, there is also a conceptual need to consider *related scientific activities* (RSA), which are 'activities that complement and extend R&D by contributing to the generation, dissemination, and application of scientific and technological knowledge' (Industry Canada 2002, 26). Federal documents enumerate the subgroupings of RSA by field of science, namely, '*natural sciences*: scientific data collection, information services, special services and studies, and education support; and *social sciences*: general purpose data collection, information, services, special services and studies, education support' (ibid.). RSA is crucially embodied in the education, training, and experience of scientific and technical personnel working on the front lines of monitoring and regulatory product approvals (Doern and Kinder 2007). In the government of Canada, S&T investments consist of the sum total of investments in R&D and RSA.

In later analysis we show the emergence of broader notions of *evidence-based* policy and regulation, where evidence includes not only the natural sciences but other forms of social science, cost-benefit and risk-management analysis, and evaluation.

A related crucial issue in science-based governance centres on the different tasks concerned with, and the number of players involved in, *pre-market as opposed to post-market regulatory science* (Doern 2007; Doern and Reed 2000; Evans 2009; Van Tassel 2009). In the former, tied closely to product assessments before they are possibly allowed into markets, the assessment of risks and benefits is a concentrated and confined process involving S&T personal inside applicant firms and inside policy-regulatory bodies. In the post-market context, where products are already in extensive use involving larger populations of users, the array of scientific and medical involvement, not to mention patients, is more akin to monitoring activity with much more complex systems and exchanges of information inherent in these processes (Health Canada 2010b).

Science-based governance as *innovation policies or strategies* is also conceptually important. It refers to government policies aimed at fostering the use of the best S&T to produce new and competitive 'first-to-market' products and new production processes, along with the innovative organizational approaches and management practices that support these activities (Conference Board of Canada 2010a, 2010b). An increasing tendency for governments in Canada and elsewhere is to subsume S&T policies within broader innovation policies in their efforts to maximize the contributions of S&T to their nation's competitiveness in the global knowledge-based economy (Doern and Levesque 2002; Doern and Stoney 2009; Kinder 2010; Phillips and Castle 2010).

Closely linked here is the core conceptual literature on *national systems of innovation* (NSI) and related concepts of *local/regional systems of innovation* and *clusters*, which conceptualize national and local S&T and innovation activity and effectiveness as products of complex *non-linear interactions* among institutions in a national or regional/spatial political economy. These include interactions among universities, corporations, governments, capital markets, systems of regulation, and informed consumerism.

Science-based governance as *public goods* or *public interest science* conceives of S&T carried out in such a way that property rights cannot or should not be appropriated by private or individual owners (Doern and Kinder 2007). Basic pre-developmental research may be characterized as a public good in this sense; so also might S&T carried out to support regulatory tasks by the state be characterized as public interest science. This conception of science-based governance is also debated in terms of the funding of government and university S&T labs and agencies, and whether they should be supported 'in the public interest' by taxpayer revenues or by user fees, which imply the existence of specific private benefits for which private users should be charged. The 'public interest' is always a contentious concept politically because it triggers diverse interpretations of democracy as well as of public policy goals.

The Analytical Framework: Three Core Elements

The strands of foundational literature, along with our early empirical research into the substance of bio-policy and bio-governance, enabled us to generate the three core elements of our analytical framework. These constitute a more manageable analytical package needed to understand, explain, and ultimately tell the biotech governance story. A brief list in each section below previews each element and some of its main features and attributes conceptually and empirically.

The State as Biotech Supporter and Biotech Regulator

Understanding the state's complex roles as both bio-supporter and bio-regulator is central to a nuanced appreciation of the changing governance of biotechnology and its three bio-realms.

On the state as biotech supporter, the following features are highlighted:

- Support as an idea expressed in changing discourse and emphasis influenced by interest-based power
- Support delivered through action and inaction and key instruments of taxation, spending, regulation, and exhortation
- Support through science and research funding and network formation

On the state as biotech regulator, the main features include:

- Regulation as overall rule-making, product approvals, compliance, and monitoring
- Regulation as networked public-private regulatory governance

And, on the interplay between these roles of the state, we examine:

- The support-regulation dichotomy model and nexus: firewalls and independent sub-arena regulation versus coordinated horizontal or joined-up overall governance

Support and promotion as an idea and activity is expressed in changing kinds and levels of discourse as well as through actions and inactions in the deployment of policy-governance instruments such as taxation, spending, regulation, and persuasion or exhortation. Regulation, as previously noted, involves an increasingly complex set of public- and private-interest-oriented rules, product assessments, compliance, and monitoring activities. The inevitable tensions and contradictions, real or perceived, by public and private interests regarding support versus regulation results in an often dense nexus of coordination activities and arenas where governance is characterized by efforts to simultaneously construct firewalls of independence and realms of horizontal coordination or joined-up governance (Bouckaert, Peters, and Verhoest 2010).

Supporting the biotechnology industry is but one example of the long-standing practice of public authorities in Canada and elsewhere assisting, encouraging, and partnering in industrial development and economic growth and in local-regional innovation and cluster development. Indeed, the National Biotechnology Advisory Committee (NBAC), the predecessor to the Canadian Biotechnology Advisory Committee (CBAC), recommended in its last major report that the minister of industry should champion biotechnology, 'recognizing that Canada's ability to adopt biotechnology and pursue its application and development will significantly determine the country's future economic status and its role in world affairs' (National Biotechnology Advisory Committee 1998, 3).

Support and promotion involves not only support for the few hundred Canadian biotechnology firms but also farmers, fishers, and other sectors of the economy. Ideas on government's role as supporter include funding further research and development, issuing intellectual property rights (patents), and encouraging the commercialization of biotechnology products. They also include marketing and securing market access around the world and streamlining biotechnology product approvals so that regulatory systems are competitive with Canada's major trading partners. By actively promoting industrial development, Leiss and Chociolko (1995, 259) claim that some in business and in government are 'risk-promoters' – with 'a direct interest in exaggerating benefits and underestimating risk,' hence interfering with the perception (and reality) of government as a neutral assessor of risk.

The concept of support, however, is rich both in theory and in practice, with several shades of meanings and possible activities, ranging from micro private interests to a larger public interest (Prince 2000). The focus of the support role by government might be a particular application or product developed by a single firm; local or clusters of biotechnology firms, such as in Saskatchewan; biotechnology as a particular sector (for instance, health care or food) or a Canadian industry; and even biotechnology as a direct contributor to the attainment of other federal, provincial, or intergovernmental public policy objectives. In other terms, the aim of support can be innovation by individual firms, regional economic development, and industrial or trade policy at the sector or national level, or some notion of quality of life and sustainable development.

Government as regulator equally has more gradations than is often acknowledged. A conventional view of regulation centres on the legitimate coercive powers of state. This definition of regulation, by emphasizing its legitimate coercive powers – its negative policing role – implies that regulation is both detached from and intrinsically antagonistic to the promotion of economic activities. Regulation, thus defined, circumscribes and restricts the behaviour of individuals and institutions, but this means excluding important areas of and roles for regulation. By contrast, social and political reality errs on the side of breadth in understanding regulation (Doern et al. 1999). If regulations are, generically, rules of behaviour backed by the sanctions of the state, then they can be, in the view of different interests, rules expressed:

- as constitutional or quasi-constitutional rules, such as the Charter of Rights and Freedoms or the 1994 Internal Trade Agreement;
- in statutes such as the Canadian Environmental Protection Act;
- in delegated legislation or 'the regs' that include eligibility rules and reporting requirements;
- as domestic or international guidelines, like Health Canada guidelines for novel foods or WHO and FAO guidelines; or
- as standards and codes, for professional ethics or research activities.

Since regulation is multifunctional, it is simplistic to characterize regulation wholly as negative and reactive, with the intent and/or effect of hindering innovation, impeding economic growth, and thereby weakening competitiveness (Conference Board of Canada 2010b). Of course, *policing* the behaviour of firms and industries is an old and – as recent global economic and banking instability and unrest starkly demonstrated – still important function of the state. Inspection, enforcement, and compliance are familiar illustrations of this function. Yet regulatory bodies in Canada also engage in *promoting* the economic welfare of firms and sectors subject to regulations and participate in *planning* by directing economic activities toward public policy aims.

As Schultz and Alexandroff (1985) point out, each of these roles has a distinctive decision-making style and a narrower or more comprehensive scope of policy goals. Depending upon the configuration of interests and power in a given sector and time, such promoting and planning may be dominated by private interest or led by public interest, reflecting different degrees of corporate autonomy and political intervention. While the philosophy of the day may create a dominating interest, whether in the form, say, of an active industrial strategy or a laissez-faire stance to markets, there are usually competing interests and conflicting objectives at any given time.

Many people, inside and outside the government and the scientific community, understandably see the relationship between the federal government's role as regulator of risks and its role as supporter of the biotechnology industry as being deeply problematic. This perspective holds that government science and/or government regulators are not always objective and impartial in the execution of their work, or in the ultimate reception given to their advice, because it is unduly influenced by bureaucratic, political, or commercial interests (Hutchings, Walters, and Haedrich 1997; Jarvis 2000; Leiss 2000).

Certain officials, a given institution, or indeed the government as a whole face performing what some in civil society and the market economy identify as incompatible policy aims and program activities. Three variants of this role conflict can be identified in relation to GM foods and to biotechnology more generally. One is that public awareness and/or governmental expectations are unclear and too poorly formulated to provide a consensus and coherent direction to policymaking and regulation. The lack of effective citizen engagement in establishing new structures and policies can contribute to this lack of consensus. Moreover, the nature of risk regulation itself, or more accurately risk-benefit regulation, is bound to generate at least some controversy regarding expectations and awareness, given differing perceptions of risk, benefits, and uncertainties, and regarding the inability to completely eliminate risk and achieve some kind of absolute notion of safety (Saner 2010).

A second kind of role conflict is that groups disagree over what the government's proper role should be, if there is one at all, in the science/regulation/support trilogy of activities. Even if there is agreement that government ought to perform all of these functions, the third version of this role conflict concerns whether a particular government department or agency, like the Canadian Food Inspection Agency (CFIA) or

Health Canada has two roles within its own mission and mandate, such as safety regulation and trade promotion, which raise honest doubts as to whether both roles can be suitably performed in a consistent manner.

In part, the concern is about the public service values of professionalism, integrity, and neutrality. More broadly, it is about our understanding of the public interest and private interests and their proper relationship. As Jarvis (2000, 316) expresses it, 'The public interest, in this case, the protection of citizens and stewardship of common resources, and private interests, determined in part by access to markets and international competitiveness, are the yin and yang of science-based regulations. These two sets of interests provide the tension that forms the basis for policy choices in this area, as they do in many areas of public policy.'

This comment is instructive in pointing out that the regulator-promoter issue is not unique to the GM food or biotechnology area, nor is it a new issue on the policy agenda of modern states. There are lessons and useful practices to be found in other quarters of government. For Jarvis, these sets of interests, while competing, are not fundamentally contradictory, and so the challenge and opportunity for governments is indeed to find a balance between them. This stance is similar to the federal government's position under the 1998 Canadian Biotechnology Strategy, which we discuss in the next chapter.

In addition to the obvious challenges associated with managing the politics of contending interests, this issue is critical for other reasons too. The nature of the separation or, conversely, the integration of regulation and promotion has implications for ministerial responsibility and organizational design. The more the separation of functions, the wider the span of controls for senior management, and the greater the need for coordination and supervisory tools.

The nature of governmental functions and their combination within an agency's mandate – rule-making, enforcement, adjudicating disputes – are also important considerations in administrative law and judicial review. Courts in Canada apply different tests for natural justice and procedural fairness, depending on the function(s) exercised by a public authority. A higher standard is applied to judicial and quasi-judicial powers than for legislative and policymaking powers. Moreover, where the tribunal members or tribunal staff exercise overlapping functions in a multi-tiered decision process, such as investigation and adjudication, courts have held that such situations may lead to a reasonable apprehension of 'institutional bias' in a substantial number of cases

(Jones and de Villars 1999). The result is that there is a real or perceived absence of independence and a sense that final decisions are predetermined.

What we can call *the regulation-support dichotomy* is a descriptive and prescriptive model of risk-benefit management and decision-making. The model holds that science-based regulation should be, and in large measure is, removed from promotional activities within government. While it may be quite legitimate for governments to promote trade and market access in regards to science and technology policies, such aims ought to be distinct from the personnel, criteria, and decision processes used for assessing biotechnology products. We are not suggesting that science and regulating are self-contained realms, or that scientific analysis and advice should be disconnected entirely from management decision-making. Rather, the intent is to ensure that determinations of the quality, safety, and efficacy of GM foods and other bio-products and processes be based primarily on scientific tests and state-of-the-art research and evidence, in compliance with statutory provisions.

Even if regulatory and promotional activities are separate at the operational and case-by-case decision-making level, they need to be coordinated and considered together at the level of policy and governmental mandates. In reality, regulation and support/promotion rendezvous in a number of ways and arenas of the federal state: in other governmental roles such as communications strategies and planning for scientific support; during major consultations with stakeholders and multiple publics; within meetings of senior coordinating committees of officials, which can bolster both regulation and promotion; during horizontal policy initiatives such as those on sustainable development; during trade negotiations and foreign policy discussions and negotiations; and as a part of annual budgetary processes.

The regulation-support nexus, therefore, is multifaceted in the number and variety of interconnections evident in the federal government. This nexus is embedded within, and shaped by, horizontal policy and managerial, legal, and constitutional systems. As a set of governmental functions, it is clear that biotechnology is structured along multiple departmental and agency lines. As a policy field, with a complex bundle of linked ideas, instruments, and interests at play, biotechnology cuts across ministerial portfolios and agencies, and their mandates, clients, and stakeholders. Likewise, it cuts across levels of government within Canada and with other nation states and international bodies. As biotechnology is a series of techniques, based in science and engineering, and not a single type of product, it has a wide scope of actual and possible applications, each with benefits and risks, which affect the mandates and authorities of several federal organizations. Each of these organizations is the custodian of a mix of values and goals, each a part of the 'public interest,' that pertain to the cultural, ethical, economic, social, scientific, and political aspects of biotechnology.

The scientific disciplines, professionals, and expertise relevant to biotechnology for food, health, and human life spread out over several government and non-governmental organizations. In the language of Ottawa central agencies, biotechnology has a high degree of 'horizontality' or interdependence as a policy field and hence in its governance. Of course, as we have emphasized in the discussion here, such horizontality is no guarantee of easy coordination, seamless coherence in policy, or, for that matter, high priority on the federal agenda as a national issue.

Networked Biotech Governance Interests and Democracy

Any political account of the forces driving or constraining bio-governance needs to deal with networked bio-governance interests and democracy – the second element in our analytical framework – the key features and attributes of which include:

- business interests in bio-food, bio-health, and bio-life;
- finance and capital interests and needs for biotech firms;
- NGO interests (consumer, health, environmental, social);
- individuals, identity, and public opinion;
- science as the quintessential network;
- government departments and agencies as interests;
- quasi-public-sector arm's-length entities as interests: universities, hospitals, foundations, and research labs and granting agencies; and
- diverse overlapping notions of democracy.

The notion of networked governance interests is central to our analysis. There clearly are interest groups and interests, and there are networks, but analytically the two descriptors must increasingly combine in biotech governance. This is because interest groups and interests are themselves networked and partnered in their actions. But also networks are, as we have seen, inherent in the nature of the sociology of science and in the nature of biotechnology, and they are increasingly being forced in ever more intricate ways by policy mandated and required levered funding. Examples of this array of networked interests appear in chapters 3, 4, 5, and 6.

When seen as a basic set of interest groups and other interests such as individual bio-firms, biotech governance embraces bio-business interest groups such as BIOTECanada and numerous bio-firms; finance, banking, and capital-providing firms; NGO interests, including consumer interests and, increasingly, individuals as citizens acting as consumers, patients, and social actors; and numerous federal departments and agencies acting in the public interest and in their own interests as promoters and regulators. This network also includes quasi-public or arm's-length entities such as universities, hospitals, and government labs engaged in bio-research as well as research-granting bodies and funding entities.

In keeping with the core notions of networked organization discussed earlier, the nature of such relationships among these bio-interests is diverse. It is not difficult to see samples of both informal networking (and alliances) and formal policy-induced or required ones in keeping with the explicit promotion of joint research, levered funding, and de facto co-regulation. Networking also occurs within the science community overall as in the nature of scientific exchange, and in peer review as the quintessential and long-established network.

When looking at *bio-business interests*, the first feature to appreciate is that the industry is awash with fast changing corporate restructuring as new sub-sectors of the industry emerge, including integrated bio-pharma companies, contract research companies, drug discovery companies, and diagnostic companies. In this restructuring context, it should be stressed that virtually all of the four hundred or so Canadian companies are new in the last decade, thus complementing the earlier, much smaller set of bio-food-focused firms.

The industry is essentially dominated by small and medium-sized firms with limited cash flow to survive the high front-end R&D development and regulatory costs; hence, the importance to these firms of financial banking interests and of risk capital. Shortages in risk capital are likely to be even greater as high volume and tailored niche products come into the pipeline (Kumar 2010). The bio-health industry argued that market forces increasingly required nationally and internationally *efficient* and effective regulatory regimes. The high proportion of international alliances among smaller and larger firms was also linked to this argument. For these and related reasons, industry sources became concerned by what they claim to be the relatively slower average approval time for new drugs in Canada. They were certainly aware as well that *effective* health and safety regulation, nationally and internationally, is also vital to their corporate prosperity. Without regulation by governments, there is no biotechnology product that is a credible product to patients and consumers.

We view *NGO interests* and related networks in three ways: as a loose coalition or assemblage of organizations, as a system of activities and beliefs rooted in core social values, and as a set of public stakeholders forming a vital part of the biotechnology policy community. Organized actors in this realm of interests include academia (colleges and universities and their research labs and institutes, in effect, quasi-governmental organizations); animal rights and welfare activists and agencies; consumer advocates and groups; environmental organizations; faith communities and religious groups; health professionals and associations; labour unions and federations; other non-governmental organizations, including registered charities and other volunteer agencies; and public research organizations. Described in this way, NGO interests closely resemble the concept of 'civil society' referred to earlier and hence to broader notions of democracy.

NGO interests, and biotechnology business interests overall, share some features. Both are more a collection of activities and processes across many different organizations and fields than a specific institution or single area of the economy or society. Most NGO interests in Canada are relatively small, with fewer than fifty employees, and many with far less staff than that, as are biotechnology firms, especially the newer bio-health firms. Moreover, many biotechnology firms have little 'vertical integration,' which means they tend to concentrate on performing only one or two functions in-house. As a result, they need to establish linkages with other organizations, through alliances and coalitions, for resources and support. The same applies to most NGO interests and organizations.

When these groups or coalitions engage in biotechnology issues and regulatory processes, they may be criticizing a government department or a particular firm or product, but they are also expressing and defending certain values and principles. Genetic applications, also often called genetic engineering, for example, disturb deeply the ethical and belief systems of many Canadians.

Different NGO organizations highlight different sets of values and principles. Among other values, environmentalists stress biological diversity and sustainable development; faith and religious groups emphasize the sanctity and dignity of human life; organized labour organizations naturally accentuate worker health and safety; and public interest research groups underline the right to know, freedom of information, and other procedural values.

The commercialization of human genetics poses basic questions of what is 'normal' in our society, in a debate that organizations of and for persons with disabilities have certainly entered, along with other groups (Caulfield 1996; Prince 2009; Reiss and Straughan 1996; Turney 1998). In addition, there is the complexity of 'interests within interests,' such as differing perspectives by physicians from the medical associations and from Health Canada or even from within a government portfolio such as Health Canada and the Public Health Agency of Canada. Thus there are changing notions of what medical power is. Doctors are still at the centre of many aspects of health-system governance but they are also increasingly networked with related health professions, with the drug industry and with health seen as an industry rather than just as a health-care sector, an 'industry' that includes health researchers in universities and in the Canadian Institutes of Health Research (CIHR) and other research institutes. This extended value diversity and differentiated system of networked power is both a policy reality and a governance test to face.

The 'density' of the NGO interest in the biotechnology policy community – the number and range of types of groups involved in a given sector – appears to vary across sectors of the Canadian economy. This variation in stakeholder engagement was initially evident in the sector consultations for the Biotechnology Strategy Renewal exercise done between 1996 and 1998. Some sectors, such as aquatics, forestry, and mining and energy identified in their consultation reports a limited number of third-sector stakeholders: university academics and scientists, and environmental and consumer public interest associations. Other sectors reported responses to their consultation documents from a much wider and deeper network of stakeholders, such as the environmental, agricultural and agri-food, and health sectors.

Consumer interests are a significant part of this large array of NGO interests. Indeed, they were a crucial presence in the initial evolution of the bio-food realm (Knoppers and Mathios 1998). Discussions of consumer interests in general and in later bio-governance periods necessarily bring out the ultimate complexity and nuanced nature of what actually is the modern consumer-citizen (Doern and Wilks 2007; Locke

1998; Middleton 1998). The consumer is best seen as being always hyphenated with some aspect of his or her role as citizen and as economic and social actor. These hyphenated roles range from the individual buyer-consumer to the environmental-consumer concerned about how goods and services are made, and the private and public goods consumer with views about balances between public and private goods.

In the bio-life realm, we will see how these extensions of citizenship and what the use and consumption of products might mean increasingly embrace a politics and bio-governance centred on the views and strengths of women's groups and coalitions, and with varied kinds of individual identity. Attention is also paid to engagement through public opinion surveys, focus groups, and Internet-based mobilization and communication.

Government departments and agencies vary when it comes to which consumer groups or institutions and also smaller subgroups tend most to engage with them and criticize them. For example, in the fair labelling practices area of the CFIA, the Centre for Science in the Public Interest has been a regular participant, as have Quebec-based groups such as the Union des consommateurs and Option consommateurs. The Consumers Council of Canada and the Canadian Consumers Association have had some, but less regular, involvement. Arguably, it is in Health Canada's institutional realm for bio-health and other mandate aspects that consumer subgroups are the most diverse and particularized. Basic subgroups here start with an overall division among patients, drug users, and special health sub-populations needing both market-based products and public services. Within and across these three categories there are any number of disease-focused groups and voluntary sector bodies dealing with bio-health and bio-life issues, as well as other issues such as AIDs, arthritis, cancer, and diabetes.

In the end, there is an explicit need to examine *government departments and agencies as interests*. Defined broadly, this set of public sector interests also includes universities, hospitals, foundations, research labs, and federal research-granting agencies such as the Networks of Centres of Excellence program, the Canadian Institutes of Health Research, and its virtual reality institutes and peer-review panels. Here there is a further array of actual primary governance structures engaged in diverse overlapping notions of networked democracy. They have public interest mandates but also have interests of their own to preserve and expand their mandates and to demonstrate success. Much of this related set of interests has been affected by overall changes in federal research and innovation policy, which in turn have changed federal government–university relations (Doern and Stoney 2009), including in the three bio-realms.

It is evident that in the field of biotechnology, governments by no means automatically march to the drum of bio-business lobbies. There is considerable inertia involved as they take into account alternative points of view from many NGOs opposed to biotechnology or insistent on appropriate public interest and stewardship standards of regulation and regulatory processes. Such pressure has certainly been present in Canada, and indeed such NGOs have been given explicit arenas for expressing their concerns, such as through the establishment and functioning of CBAC, until its demise in 2007. Moreover, as we discuss further below, the regulators themselves, armed with complex statutory mandates and duties, have never simply done the bidding of the industry.

Science-Based and Precautionary Governance

Science-based and precautionary governance is the third element in our analytical framework. In science-based governance, key attributes and features include:

- science-based or sound science in particular kinds of science, cast as R&D, related science activities (RSA), evidence-based policy and regulation, and innovation;
- science-based governance interacting with intellectual property 'invention-based' governance;
- science-based governance as designed technology-assessment processes; and
- science-based governance as influenced by networked science communities.

In precautionary governance, the main features include:

- the precautionary principle and scientific uncertainty as a governance norm;
- precaution inherent in decision stages of risk assessment, risk management, and risk communication;
- precaution inherent in risk-benefit assessments at the level of rule-making, pre-market product assessment/approval, and risk-benefit monitoring in post-market contexts; and

- precaution as an ill-defined 'resultant' of complex governance within a bio-realm and across bio-realms.

This must be discussed with a good analytical respect for the diverse meanings inherent in science-based government, including notions of R&D, related RSA, innovation policies, public goods, and linear and non-linear conceptions of science and technology pursued by regulators, government labs, firms, universities, and foundations.

Aspects of transformational technologies, to be discussed in chapter 3, are also embedded in our conceptions of science-based governance, along with the trade-related discussion in chapter 2 of the central role and defence of sound science. Our discussion on networked governance underscores the roles of the science community as being preeminently network-based and thus is also inherent in what must be empirically looked for in science-based government.

Because of the intrinsic contested nature and growing importance of intellectual property, in biotechnology, particularly patents, we need to look closely at such 'invention-based' governance. Patent policy and patent approvals have key scientific and technical dimensions and are tied closely to policy goals regarding innovation. They relate increasingly to the limits and abuses of patenting, to 'patent thickets' (complex multi-patent compositions of some products) and hence to arguments about public versus private ownership of such rights, given other public-interest and invention-dissemination goals central to patent policy and to how innovation actually works (Castle 2009; de Beer and Guaranga 2011).

The pairing of science-based governance with precautionary governance is necessary in several analytical senses. First, the precautionary principle emerged as an explicit governance norm in environmental policy, national and international, partly to counter trade-related notions of sound science (Dickson and Cooney 2005; O'Riordan and Cameron 1994; Saner 2002). As a governing norm and procedure, the precautionary principle stresses the need to respect the realities of scientific uncertainty and the need when warranted not to proceed with some products or processes unless one can have much higher confidence in evidenced-based standards of science and knowledge overall. Precaution is not itself the opposite of science-based governance. It is not wholly the same either, even though advocates of the precautionary principle insist on the use of science as well.

The second sense in which precaution inherently emerges is in the core trio of processes that accompany risk-benefit decision processes.

These refer to the regulator's applied decision stages of risk assessment, risk management, and risk communication whose operation are then seen by some agencies as notions of actual precautionary behaviour regarding new rules that might be needed, but especially regarding new products and processes under review by regulators. There are also notions of precaution integral to the different decision dynamics of pre-market product assessments/approvals and post-market monitoring where, in the latter, typically vastly larger sets of networked players and users are involved.

A further sense in which precaution emerges is in the notion of technology assessment, discussed in the next chapter. Advocates of formal technology assessment processes usually see the need for more complex and democratic forms of engagement. Typically, these are precautionary in the sense that a wider range of values and viewpoints should be engaged and that such processes take time and should take the time needed, democratically and practically.

Finally, and particularly in the context of our governance focus, it is essential to look for forms of precaution that are not designed as such but rather are the result of complexity within realms, and also among the three bio-realms. Bio-governance has grown ever more complex and multi-layered, and this too – through bio-regime political, economic, and administrative inertia and transaction costs – can induce precaution in possibly unintended and even unrecognized ways.

Conclusions

From three quite broad strands of literature, we have set out the conceptual foundations underpinning the analysis, which suggest to us the necessity to examine several areas and depths of inquiry about governance *and* biotechnology in substantive terms in Canada over the last three decades, across and within the three bio-realms.

The state as biotech supporter versus biotech regulator, networked bio-governance interests and democracy, and science-based and precautionary governance constitute the three main elements in our analytical framework, each with several attributes and features that the reader needs to be aware of as the narrative proceeds. There is no escaping the fact that bio-governance is complex governance and is replete with overlapping and reinforcing features within and across these three elements; contradictions and conflicts among them are also at play.

In the bio-food, bio-health, and bio-life realms, regulatory aspects loom larger in the narrative, complemented by attention to some support governance norms built in, as well to regulatory processes, values, and laws. The nature of their different networked interests and concepts of democracy and their varied bases in science-based and precautionary governance emerge clearly from these accounts. So, too, do the different core politics these three bio-realms exhibit and have to manage in swiftly changing arenas of democracy and national and global circumstances.

2 National and International Biotechnology Policy in Liberal and Conservative Government Eras

Introduction

Biotechnology policy[1] is inextricably a set of linked and evolving national and international policies, with each influencing the other domain in timing and content. Canada is a part of international negotiations in biotech-related policies on trade, patents, and health, and, in most cases, has agreed with and signed on to the policies adopted. This chapter surveys the major biotech policy statements, strategies, and changes in the last two decades of the twentieth century and the first decade of the twenty-first century. The analysis shows important features of partisan political party aspects of biotechnology policy and its links to representative Cabinet-parliamentary democracy. These partisan dimensions show continuities across Liberal and Conservative governments and prime ministerial eras, as well as some differences, as biotechnology policy emerged and expanded in scope, nationally and globally.

Aspects of multi-level governance and multi-level regulation also appear in this historical narrative, as new forms of coordination and conflict materialized, and as the biotech governance regime became more complex, embracing both direct biotechnology policies and indirect ones. The chapter also offers some further observations on the core politics of global biotech policy, especially that between the European Union versus the United States and also the continuous impact of U.S.-Canada relations.

Biotechnology policy includes policies where biotechnologies are referred to directly in Canadian and international policy discourse. It also emerges more indirectly when other existing Canadian and international policies, policy fields, and their related discourse are said to apply to biotechnologies along with any number of non-biotechnology products and processes. Such policy fields include trade, health, environment, research, innovation, and patent policies. As biotechnology policy expands across the varied bio-realms, government departments and agencies, and time, it includes a further related layering of normative content expressed as principles such as stewardship and citizen engagement (Abergel and Barrett 2002; Commission of Inquiry on the Blood System of Canada 1997; Doern 2003; Kuyek 2002).

To provide a contextual historical overview of biotechnology policy, this chapter proceeds through the time periods of the main federal prime ministerial and governing political party and partisan eras: the Trudeau Liberals in the early 1980s, the Mulroney Conservatives in the 1984 to 1992 period, the Chrétien and then Martin Liberals in the 1993 to 2005 period, and finally the Harper Conservatives from 2006 to the present. In general, direct bio-policy in Canada has operated in the middle and lower-ranked realms of political priorities, in that biotech policy has received few and brief mentions in the pinnacle federal agenda-setting events of Speeches from the Throne and Budget Speeches. This is an important fact and argument in the overall story to be told in this book. Our initial historical account will be complemented in later chapters with more detailed histories and analysis as our focus shifts more fully into a bio-*governance* framework in the three bio-realms.

While the chapter provides a broadly chronological account of bio-policies, the actual underlying bio-technologies in bio-food, bio-health, and bio-life were not always occurring in orderly chronological ways. They did not all lead immediately to recognition and understanding by governments or clearly to their own particular policy statements. Particular bio-products, processes, and research findings were more random, multi-dimensional, and overlapping, surfacing throughout the three decades being examined and even before then.

In the bio-life realm, for example, the conceptions of, and debates about, reproductive technologies stretched across two decades until laws and relevant regulatory agencies emerged, and even then with only partial and still contested governing structures. Biotechnologies were a part of reproductive technologies, the latter including other technologies as well. Similarly, in the bio-health realm, there was a burst of genome-related products and processes after the human genome mapping was announced in 2000, but plainly other DNA-related technologies came earlier. Many biotechnologies and products still do not show up in recognizable policy statements or broader governance structures.

Biotechnology Policy in the Trudeau and Mulroney Eras: Early Support and Major Trade and Patent Policy Impacts

The foremost way in which biotechnology policy emerged in the 1980s was its explicit recognition as an important field and industry by the Ministry of State for Science and Technology (1980) and then in the gradual fashioning of a bio-food-centred biotechnology regulatory system in response both to international developments, especially in the United States, and to the development of bio-food products. This led to the establishment of a Federal Regulatory Framework for Biotechnology. Under this framework, there was and still is no *single* biotechnology regulator, although there now are offices of biotechnology within both Health Canada and the CFIA. As an alternative to forming a single regulator, a framework of principles was developed to guide the several regulatory bodies and departments that were being called upon to assess biotechnology products. These included Environment Canada as well, in limited ways.

A second way in which some aspects of biotechnology policy emerged was through the establishment in 1989 by the Mulroney Conservative government of the Royal Commission on New Reproductive Technologies. The main impetus for the commission, which reported in 1993, came from a coalition of feminist professional women concerned about already available reproductive technologies and their impacts on women and the rights of women (Miller Chenier 1994). Biotechnology, while not explicit in its mandate, became a part of its concerns, as the research into the genetic aspects such as prenatal screening, gene therapy, and genetic alterations evolved.

A third way in which biotechnology emerged as a bio-policy issue in the 1980s was through an initial 1983 Trudeau-era National Biotechnology Strategy. It was essentially and unambiguously an effort to support and promote R&D, investment, and private market acceptance of this new technology, accompanied by the establishment of the NBAC, whose early work helped pave the way for the later work of the Canadian Biotechnology Advisory Committee. While NBAC was established in the context of a fairly pro-biotech policy strategy, its research and reports began to include concerns about citizen consultation and broader economic, health, and stewardship values as part of more balanced approaches to biotechnology (Doern and Sheehy 1999).

The Mulroney era was decisive for its launching of free trade both in the Canada-U.S. Free Trade Agreement and in the later North American Free Trade Agreement (NAFTA). Interaction between these trade agreements also helped produce the even larger Uruguay Round global trade agreement that led to the establishment of the World Trade Organization (WTO) and also produced policies on trade dispute settlement, science-based policy, and patents with major biotechnology policy implications. Ultimately, these extend into the Chrétien era but are best discussed in some detail as a policy legacy of the Mulroney era.

This emerging international regime for biotechnology became a part of the larger WTO trade policy system, including the WTO-centred Trade-Related Intellectual Property System (TRIPS) agreement (Sell 2010). Biotechnology is simply one unnamed embedded product or process that can be traded internationally and hence might be caught up in the larger WTO trade rules and processes. Three key WTO aspects are central to biotechnology policy: WTO dispute settlement, its science-based norms/rules, and TRIPS.

First, the dispute-settlement system under WTO has more regulatory teeth than the earlier General Agreement on Tariffs and Trade (GATT) processes and is intended to work more expeditiously (Petersman and Marceau 1997; Trebilcock and Howse 1995). Dispute resolutions have a strict time limit established for the conclusion of the process, and a single member country is prevented from blocking the adoption of reports of trade-dispute panels, or, on appeal, of appellate bodies. The WTO is also more proactive, since it has a trade-policy review mechanism in which it has an independent investigative authority to initiate rotating country-by-country reviews of international and domestic policies that might adversely affect trade relationships among countries. At the insistence of the United States and the European Union (and with strong Canadian support), these core mechanisms were intended to entrench liberalized trade and minimize obstacles to trade, including trade in biotech products.

A second WTO feature centres on ideas about science in international trade (Lal Das 1999; Trebilcock and Howse 1995). Core WTO and other trade provisions (e.g., NAFTA) are fixed firmly on the importance of objective and transparent science (in short, credible scientific evidence) to underpin international regulation and standard-setting in health, safety, and environmental matters. Science-based regulation is seen as crucial to ensuring that health, safety, and environmental rules do not distort trade or become the new guise for protectionism (Doern and Reed 2000; Vogel 1995). There has always been a tension in trade agreement negotiations between environmental criteria where the precautionary principle is advocated (more on this below) and trade or economic criteria where a higher threshold of science or scientific evidence is advocated or agreed to in negotiations.

Trade-related ideas about science have emerged in debates about eco-labelling and in the issue of whether campaigns led by environmentalists for competitive informal standard-setting, such as a consumer boycott, undermine official science-based standards sanctioned through trade agreements. Such informal efforts are seen by some as a threat to democratic, formalized, international, science-based regulation and harmonization whereas, clearly many other groups and environmentalist coalitions see it as a necessary democratic counterweight to the institutionalized or official trade system.

A further aspect of the WTO, associated with other international intellectual property norms and institutions as well, is TRIPS. Under the earlier pre-WTO General Agreement on Tariffs and Trade, the international regime for intellectual property (IP) fell well short of a harmonized regime (Doern 1999; Doremus 1996; Drahos 1996, 1997; Sell 1998, 2010). Indeed, IP issues were largely outside the GATT purview. Moreover, international IP organizations such as the World Intellectual Property Organization (WIPO) did not contain a formal court-like process for dispute resolution. It regularly reported on disputes but had no GATT-like panel process to resolve disputes (Doern 1999; Trebilcock and Howse 1995).

The Uruguay Final Round Act included for the first time a comprehensive agreement on TRIPs that seeks to balance the conflicting values inherent in IP and between developed and developing countries (Bhat 1996). It greatly strengthened the role of the WTO but it also established a new body, the Council on TRIPS, and mechanisms to help developing countries get ready for the new stricter regime.

During the Uruguay Round, the issue of mandates in IP between the WIPO and the proposed World Trade Organization generated considerable dispute (Doern 1999). The developing countries preferred the WIPO as the lead institution because it had facilitated diverse IP policies and institutions in developing countries. The United States and Europe, but especially the former, preferred a stronger WTO mandate because it wanted better dispute settlement and enforcement of harmonized IP rights, especially regarding key developing countries whose IP regimes were either weak in law or weak in their implementation. As a result, the Uruguay Final Round Act included for the first time not only the WTO itself but also a comprehensive agreement on TRIPS that, in principle, seeks to balance the conflicting values inherent in IP and between developed and developing countries (Marcellin 2010; Sell 2010).

Central to WTO-TRIPS is also the achievement in the Uruguay Round of a harmonized patent of twenty years' duration from filing date. The length of patent protection is an issue replete with economic and political calculation and pressure. A key question is 'why twenty years?' and another is 'why one period for all industrial sectors or kinds of invention?'

Patents produce a temporary monopoly to reward intellectual effort and ingenuity. But simple economic logic suggests that these periods of protection ought to vary greatly by field or sector depending on varying cost structures, investments, and payback periods. This also suggests that countries would have different views about what kinds of protection across sectors would make the most sense, given their national state of development and strategies for development. Thus the market economics underlying patent protection suggests the suitability of many periods of protection, and these periods could also change over time.

However, as is the general case in democratic affairs, the political and institutional logic of patent policy differs somewhat from the market logic. For major economic actors, the basic logic is simply the longer the protection period the better. This view is driven by firms such as those in the national and global pharmaceutical and biotechnology industries who sought out and achieved maximum effective protection. Their desire for maximum periods is driven by factors such as high upfront costs in R&D and in obtaining ever-lengthening drug approval processes by other government regulators in several countries. In the 1980s they saw their *effective* protection being reduced and sought change in national laws and trade regimes.

The 'longer is better' logic was also the driving force behind the approach of United States, and later the European Union, in successive trade negotiations. U.S. power was crucial in this regard, in that the Americans saw IP as increasingly crucial for their domestic economic development in 'new economy' industries such as biotechnology and related it both to developing countries with weaker regimes on patents and to fellow-developed countries such as Canada, which it pressured to change its patent laws as well (Trebilcock and Howse 1995). In Canada, this pressure focused on Canada's preferences given to generic drug manufacturers and was brought to bear before and during successive FTA, NAFTA, and GATT negotiations (Doern and Sharaput 2000).

In the late 1980s and early 1990s, in particular, there were few if any effective counter-pressures from those interests/countries making the counter-arguments. Developing countries mounted some counter-pressure but were eventually worn down by more powerful forces. Consumers in some overall sense had a vested interest in less monopolistic practices but, at national and certainly at international levels, they were a weak, diffused, and virtually voiceless interest. Perhaps the only exception was in the health sector, where health ministries and NGOs were often a surrogate representative of consumer, patient, and disease-specific interests.

Another reason for longer IP protection periods and for one long period comes from the political and administrative logic of trade negotiations and implementation. It was simply easier to agree on one such longish period because it would be easier to implement. Moreover, it would preserve the notion that IP was indeed an area of real *framework* law that applied across the economies of member states and did not constitute a form of sector-specific 'industrial' policy, which it would be if many sectoral-based periods of protection were possible.

While the TRIPS regime dealt with IP as a whole, with biotechnology nominally a part of the backdrop, pharmaceutical industries and related biotechnology firms were the core lobby in the United States pushing for the new regime, and in the 1990s they largely succeeded in getting their wishes adopted (Sell 1998).

A further area of IP patent policy is more overtly biotechnology-specific – the patenting of higher life forms (as opposed to microbial life forms, which are patentable in many countries) and the broader economic and ethical issues inherent in regulating biotechnology. American law allows such patenting, whereas EU and Canadian law does not or is only considering the need for special procedures. Debate here turns on what constitutes an *invention* and the degree to which the manufacture or the composition of matter was under the control of the inventor as opposed to control by the laws of nature. These concerns raise serious ethical issues about patenting life forms, and similar issues attend other genetic testing products and bio-life processes.

Biotechnology Policy in the Chrétien Liberal Era: Towards Better Balance amidst Extended Trade and Precautionary Norms

The foundational 1980s bio-policies were updated somewhat in 1993 by the revised Federal Regulatory Framework for Biotechnology, in effect a by-product of pressure and thinking in the late Mulroney Conservative years but supported in the Chrétien Liberal era. It provided overarching principles for the functioning of the federal biotechnology regulatory system (Doern and Sheehy 1999; Industry Canada 1998). Developed in the light of both interdepartmental and stakeholder consultation with diverse interests, the amended federal framework centred on six principles for the regulation of biotechnology:

- Maintain Canada's high standards for protecting the human health of Canadians and the environment
- Use existing laws and regulatory departments to avoid duplication
- Develop clear guidelines for evaluating biotechnology products that are in harmony with national priorities and international standards
- Provide a sound, scientific knowledge base on which to assess risk and evaluate products
- Ensure that the development and enforcement of Canadian biotechnology regulations are open and include consultation
- Contribute to the prosperity and well-being of Canadians by fostering a favourable climate for investment, development, innovation, and the adoption of sustainable Canadian biotechnology products and process (Industry Canada 1998, 12).

In various ways the six principles are intended to reflect a reasonably balanced approach between ensuring health and environmental protection broadly speaking and fostering the development and practical benefits of biotechnology products and processes, including Canadian economic competitiveness in this sector. The principles also reflect Canada's international commitments (see more below) under the United Nations Commission on Sustainable Development, the United Nations Convention on Biological Diversity, and the World Trade Organization and NAFTA (Phillips and Wolfe 2001).

This framework was then replaced in 1998 by the Canadian Biotechnology Strategy (CBS), the focus of which was much broader. The CBS was intended to 'support the responsible development, application, and export of biotechnology products and services' balanced within the context of 'social and ethical considerations' (Industry Canada 1998, 1). The CBS set out a policy framework consisting of a vision, guiding principles, and goals that reflect biotechnology's impor- tance both to the economy and to Canada's quality of life. Ten themes 'for concerted action' were identified to be implemented in partnership with stakeholders such as the provinces, industry, academia, citizens, environmental groups, and other interests.

Another key feature of the 1998 CBS was the establishment of the Canadian Biotechnology Advisory Committee as a replacement for NBAC. CBAC was established as an expert panel to advise ministers on the 'ethical, social, economic, scientific, regulatory and environmen- tal and health aspects of biotechnology' (Industry Canada 1998, 1). The CBAC had no role on specific regulatory decisions, though its policy advisory role included serving as a forum to give Canadians a voice in an 'open and transparent dialogue on biotechnology issues' (ibid.). Also established at that time was the Canadian Biotechnology Secretariat), with a mandate mainly to service the committee of assistant deputy ministers charged with coordinating the CBS, through their respective ministers, and also to function as the secretariat for CBAC.

In 1999 the Canadian Biotechnology Strategy was amended to estab- lish three strategic policy directions dubbed 'pillars': stewardship, benefits/innovation, and citizen engagement. At this point in federal discussion, the stewardship pillar referred to 'ensuring effective stew- ardship of biotechnology in the areas of health, safety and the environ- ment' (Canadian Biotechnology Secretariat 2002).

A further way bio-policy has gained a greater profile on the national stage and international scene is being central to more focused policy concerns or particular regulatory *controversies*. These concerns ranged from global scientific issues such as the cloning of Dolly the sheep, gene prospecting in many countries and its links to biodiversity, and the vast human genome research project (Appleyard 1999; Genome Canada 1999; Grace 1997; Mironesco 1998; Rifkin 1998; Shiva 1997). The genome project, which Canada joined, was certainly a key factor in the Chrétien Liberals' policy to create Genome Canada as an arm's-length network- centred research support agency.

Controversy in the Canadian context also includes specific products, such as Canada's debate over the regulation of recombinant bovine somatotropin (rbST) (MacDonald 2000; Mills 2002). This dispute, to be discussed further in chapter 3, began over a single milk product in the bio-regulatory system, but opposition to its possible approval led to a reference and much larger public examination in a parliamentary committee.

Furthermore, while the final report of the Royal Commission on New Reproductive Technologies was published in 1993, key legislation on these issues did not emerge until 2004 when the Martin Liberals succeeded in having Parliament pass the Assisted Human Reproduction Act, and it was not until 2006 that, under the Harper Conservatives, Assisted Human Reproduction Canada was established as a new agency to administer the AHR Act. During this period, these policies and agencies were being challenged continually by the Bloc Québécois in Parliament. But more crucially, they were successfully challenged constitutionally in the courts by Quebec on the grounds that the federal use of criminal law powers is ultra vires in health matters – a view mainly accepted by the Supreme Court of Canada in its December 2010 ruling (see chapter 6).

In addition to these policy principles enunciated in the 1980s and 1990s, the federal biotechnology product-assessment processes were (and are) also governed by *generally accepted approaches* that Canadian and international regulators have developed for the regulation of novel foods. These include the view that safety assessment is on the final product and that it is based initially on a comparison of the modified organism to those of its traditional counterpart, where such exists (the concept of *substantial equivalence*). If such equivalence does not exist, then broader assessments will be necessary (Canadian Biotechnology Advisory Committee 2002).

Overall regulation is also governed by principles regarding phases of decision-making, which encompass risk assessment, risk management, and risk communication. Several federal science-based regulatory agencies give expression to these principles of risk (Doern and Reed 2000; Health Canada 2010a; Saner 2010). *Risk assessment* involves scientific analysis of the likely severity of adverse health and environmental effects, the probability of its occurring, the size of the population at risk, and other related factors. *Risk management* involves analysis and positive actions and steps to eliminate, reduce, manage, or avoid risks. Such actions are determined by statutory responsibilities, commitments, and partnerships, and by assessing public health or other benefits relative to risks. The ability to manage risks is also partly a function of available resources including staff, expertise, and money. *Risk communication* involves purposeful exchanges of information about health or environmental risks between regulators and Canadians.

Other important manifestations of biotech policy during this period came in the form of new S&T granting and funding mechanisms and agencies, beginning with the Networks of Centres of Excellence (NCE) program involving the three main federal granting bodies. It also included the establishment in 1997 of the Canada Foundation for Innovation (CFI). Both of these, as we shall see in later chapters, have had significant biotech funding impacts and related network-based implications for governance.

Interwoven with these biotech-related policies were also international developments, and policies that regulate and approve new pharmaceutical and reproductive technologies and food products. At its core, this cluster of policies proceeds from a series of *national* health and safety regulatory bodies in several countries rather than from a fully developed international approval process as such. At the same time, however, WTO and other *international* trade-related dispute-settlement processes were becoming more influential for biotechnology policy.

The WTO Agreement on the Application of Sanitary and Phytosanitary Measures (the SPS agreement) places disciplines on national regulation and enforcement of the protection of human, animal, or plant life or health from risks arising from animal or plant pests or diseases, food additives, or contaminants (Lal Das 1999). The intent here is to prevent the use of SPS measures from becoming disguised restrictions on trade, while safeguarding each country's right to protect health and safety. The crucial principles of SPS measures are: avoiding unnecessary obstacles to trade; basing measures on scientific principles, scientific evidence, and risk assessment; harmonizing regulatory measures among nations; and the above-mentioned equivalency concept; as well as provisions regarding transparency.

Another international biotech-related policy element is the CODEX Alimentarius Commission, a long-standing organization set up by the Food and Agriculture Organization (FAO) and the WHO that develops food standards, guidelines, and codes of practice. In 1989, it decided to evaluate applications of biotechnology under the existing CODEX system, which meant that no separate committee was established.

In both SPS and CODEX aspects as a whole, the intent was that biotechnology was not singled out. It was just another product that, like others, had to be looked at using sound science-based regulation. The real politics of this aspect of policy and international regime construction is also found in the nature of national regulators, or rather sets of national sectoral regulators. Canada is a useful example here, as we will see in detail in chapter 4 and elsewhere.

Establishment of the European Medicines Agency in 1995 is also noteworthy, its existence propelled by the high volume of new drugs and bio-health products. The EMA is based on policies designed to ensure that smaller and even medium-sized EU member countries could take advantage of the greater scale and capacity of scientific assessment capacity in a single Europe-wide agency and not have to reproduce it themselves for their own smaller market (Abraham and Lewis 2000, 2003; Vogel 1998).

A decade later, the European Food Safety Authority (EFSA) was established with inevitable linkages and gaps in relation to the EMA (Vos and Permanand 2009). A fundamental policy feature is that the risk-assessment function was made much more transparent and was assigned overall to the EFSA, but without its advice and reports being binding on the EU's now twenty-seven member states.

As regards technologies related to human life, the European Group on Ethics in Science and New Technologies appointed by the European Commission identified in 2002 a number of international directives from bodies such as the World Intellectual Property Organization, the World Trade Organization with regard to its annex on TRIPS, the United Nations, and the Universal Declaration on the Human Genome and Human Rights (European Group on Ethics 2002).

Negotiation of the bio-safety protocol to the UN Convention on Biological Diversity is a further element of international bio-policy. The crucial event and process here starts with the forging of the convention at and after the 1992 Rio Earth Summit (Purdue 1995). The process became an intense economic and political issue between the developed and developing states of the world. In the early 1980s, developing countries had been able, through the Food and Agriculture Organization), to obtain an undertaking that said that all seeds are a common heritage. These included 'inbred elite lines used for breeding by seed companies' (ibid., 101).

By the late 1980s, developing countries had countered this with an agreed FAO interpretation that intellectual property rights were not incompatible with the earlier undertaking. Developing countries shifted their positions somewhat as the Rio Earth Summit approached, seeing seeds and plants as natural resources over which nations had sovereign authority. The resulting convention speaks of the need for adequate and effective IP protection and, in the view of many critics, ensures that the WTO-TRIPs agreement will prevail over the biodiversity agreement.

In November 1995, parties to the convention began work on a draft protocol on biosafety that focused specifically on the trans-boundary movement of any living modified organism (LMO) resulting from modern biotechnology that might have adverse effects on the conservation and sustainable use of biological diversity. The negotiating mandate pointedly excluded issues related to the safety of genetically modified food. Its focus, rather, was on the trans-boundary movement of *living* modified organisms – that is, those capable of reproduction and not the non-living products derived from them.

Several meetings and negotiations culminated in the adoption of a consensus text in Montreal in January 2000. In the final phases of negotiations, five negotiating groups emerged: the 'like-minded group' composed of most developing countries; the 'Miami group' composed of six major agricultural exporters (Argentina, Australia, Canada, Chile, the United States, and Uruguay); the European Union; the 'compromise group' (Japan, Korea, Mexico, New Zealand, Norway, Singapore, and Switzerland); and Eastern Europe, Russia, and other countries of the former Soviet Union.

Important features of the protocol are as follows (Environment Canada 2000; UNEP 2000; U.S. Embassy 2000). First, the scope of the protocol is limited in that it applies to the trans-boundary movement, transit, handling, and use of all living modified organisms that may have adverse effects on the conservation and sustainable use of biological diversity, also taking into account risks to human health. The human health reference is interpreted as health effects from environmental and occupational exposure and those resulting from an adverse impact on biodiversity. As mentioned, food safety is not addressed. Pharmaceuticals for humans addressed by other relevant international agreements or organizations are exempted, and only some of the provisions of the protocol apply to LMOs in transit or in 'contained use.'

A second feature is the application of Advance Informed Agreement procedures to imports of LMOs. These apply to the first trans-boundary movement of LMOs destined for deliberate release into the environment, such as seed, saplings, fish, and micro-organisms for bioremediation. Decisions on import are to be taken on the basis of risk assessments and pursuant to specified procedures and time frames. A signature country's failure to acknowledge receipt of a notification or to communicate its decision on import within the prescribed time frames does not imply consent to the import.

A third feature of the protocol deals with advance information on domestic approvals of LMOs destined for food, feed, and processing 'bulk commodities.' This advance information would allow countries to determine and advise what regulatory requirements, if any, would apply to the first import of such LMOs well before these LMOs enter into international trade. Another provision ensures that the documentation for all such shipments must clearly indicate that the shipment 'may contain' LMOs that are not planned for intentional introduction into the environment and a contact point for further information.

Crucially, in addition, the preamble to the protocol recognizes that trade and environment agreements should be mutually supportive with a view to achieving sustainable development as an overriding policy objective. It emphasizes that the protocol should not be interpreted as implying a change in the rights and obligations of a signatory under any existing international agreements. The protocol incorporates the dispute-settlement provisions of the Biodiversity Convention, but parties preserve their right to have trade disputes arising out of a country's implementation of the protocol's provisions resolved in the WTO.

The protocol furthermore contains references to the *precautionary approach*. Principle 15 of the Rio Declaration approved at the 1992 UN Conference on Environment and Development is referred to in the preamble and in the objectives article. The precautionary approach is also 'operationalized' in an effort to ensure that the application of the precautionary approach by a party of import is understood to be linked to a science-based process for taking import decisions.

As a central paradigm in contemporary science policy, the precautionary approach has wide currency and strategic importance in debates about biotechnology for a number of reasons (Stirling 1999). In general terms, the precautionary principle implies that action should be taken by policymakers and regulators to prevent environmental damage, even if there is uncertainty regarding its possible cause and possible extent (Connelly and Smith 1999).

According to this view, 'the environment should not be left to show harm before protective action is taken; scientific uncertainty should not be used as a justification to delay measures which protect the environment' (Jordan and O'Riordan 1995, 59). One implication of this view is that one does not need complete scientific proof to practise the principle. Jordon and O'Riordan claim the setting for the use of the precautionary principle might be more propitious because of a broad-scale critique about science in environmental policy and regulation. They argue that the 'science of assimilative capacity, predictive modeling and compensatory investment to offset the loss of ecological resilience is being challenged' (61). This and other criticisms have led to the view that 'at the very least, science should evolve into a more applied, interdisciplinary format for coping with environmental threats, and that it should be seen as a tool for a more open and participatory culture of decision taking' (ibid.).

The precautionary principle aims, to some degree, to shift the onus for the protection of human and environmental health and safety back onto those whose actions are imposing social costs, quite apart from the science that may be involved. Efforts by the European Union to devise guidelines on the application of the precautionary principle are also of particular interest and importance (Smith and Halliwell 1999). These derive partly from EU or community environmental law.

EU guidelines suggest that, as an approach to risk management, the precautionary principle applies to risks for future generations as well as to present risks, linking the principle to the paradigm of *sustainable development* as articulated by the Brundtland report (World Commission on Environment and Development 1987). Because of considerable public opposition to biotechnology in Europe, especially reflected in the European Parliament regarding biotechnology in food, the precautionary principle found its way into several parts of the biosafety protocol.

Following adoption of the protocol, American officials drew particular attention to why the United States believed that the protocol included the precautionary principle but also reined it in. U.S. Assistant Secretary of State David Sandalow first stressed that 'the biosafety protocol is the first international agreement to expressly recognize the potential benefits of modern biotechnology' (U.S. Embassy 2000, 1). Second, he pointed out that the protocol 'has ensured ... that world food trade will not be disrupted.' Third, he remarked that nothing in the protocol changes the requirement under the WTO to regulate 'on the basis of sound science' and that the protocol is consistent 'with the long-standing U.S. view that the precautionary approach should be part of a science-based decision-making process, not a substitute for that process.' Quite pointedly, Sandalow also emphasized that 'the protocol does not include a much more stringent requirement sought by the European Union and others that every individual strain or a bioengineered product be identified, in every individual shipment' (2).

Biotechnology Policy in the Chrétien, Martin, and Harper Period since 2000: Towards a Genome-Centred Bio-Economy?

In the first decade of the twenty-first century, changes in biotechnology policy and discourse were evident in the Chrétien and Martin Liberal governments and in the Harper Conservative government. These changes were aimed at revising and updating the 1998 CBS through extended biotech policies or through budgetary decisions.

A 2003–4 internal discussion focused initially on developing a blueprint for biotechnology (Canadian Biotechnology Secretariat 2004). It was informally agreed to by the committee of ADMs that deals with biotechnology under the 1998 CBS. The blueprint suggested a framework that would position Canada as a world leader in biotechnology and its applications. Its fundamental objective was to accelerate the commercialization of Canadian biotechnology research for the social, environmental, and economic benefit of Canadians. The blueprint centred on getting ready for the larger bio-economy, integrating an economic agenda with effective stewardship and promoting a comprehensive systems approach with a global approach in federal decision-making.

The blueprint stalled politically for a variety of reasons, including the presence of new senior officials in the process, the absence of ministerial interest and focus, given other priorities, the need to marry any strategy with intervening policy initiatives such as the federal government's 'smart regulation' initiatives of 2004–5, and differences of view about what biotechnology stewardship actually meant. Such differences, as we will see, have always been a part of the debate and emerged time and again.

A further 2004 bio-policy-related change came in the form of Canadian legislation to create Canada's Access to Medicine's Regime achieved through amendments to the Patent Act. This change in public health access to patented lifesaving drugs, including bio-drugs, emerged through changes to the WTO-TRIPS agreement, which were intended to create greater access to medicines by WTO member countries with insufficient or no manufacturing capacity in the pharmaceutical sector, which would allow them to make effective use of the compulsory licensing allowed under TRIPS.

Under the provisions of TRIPS (Article 31), compulsory licensing or government use of patents is allowed without the authorization of the patent owner. One of the conditions under which this can occur is when such use is predominantly for the supply of the domestic market. However, TRIPS also prevents WTO members with manufacturing capacity from issuing compulsory licences 'authorizing the manufacture of lower-cost, generic versions of patented medicines for export to countries with little or no such capacity.' In 2003, WTO members agreed to a waiver provision regarding this provision whose purpose was to 'facilitate developing and least-developed countries' access to less expensive medicines needed to treat HIV/AIDS, tuberculosis, malaria and other epidemics.'

Canada was the first country to announce that it would implement this waiver, and in May 2004 Canada's legislative framework was given parliamentary approval, and a year later its regulatory provisions came into force. Nonetheless, since then, the Canadian Access to Medicines Regime has not resulted in the export of any eligible pharmaceutical products to eligible importing countries.

By 2005, the federal bio-policy discussion and discourse had extended to a much broader notion of biotechnology *stewardship* anchoring it to a *life-cycle approach*, beginning with research and development and leading through distribution, processing, manufacture, sale, and use, and to its eventual disposal or recycling back into further research (Industry Canada 2005). The latest format for a federal biotechnology stewardship framework effectively combines the earlier three pillars and encompasses three goals:

- to advance prosperity;
- to ensure safety and security; and
- to project Canadian values here and abroad.

Supporting these three goals are several principles: innovation, accessibility, risk management, transparency, respect, and international citizenship. In combination, these layered goals and principles involve the federal government in roles as innovator, promoter, regulator, manager, user, educator, and social policy steward (Industry Canada 2005).

A further bio-policy addition was the establishment within the biotechnology framework of an International Development and Cooperation Working Group (IDCWG). International development goals had not been explicitly a part of the 1998 strategy. Thus the IDCWG became an effort to better integrate and implement a multi-departmental effort on biotechnology and related matters as guided by federal international development policies and through discussion involving federal departments such as CIDA and other players, many of whom were not used to dealing with each other regularly on bio-policy matters.

Among the policies driving the exercise was Canada's adherence to the globally agreed Millennium Development Goals (Rath 2004). Prime Minister Paul Martin's 2004 post–Throne Speech commitment on international development was also pivotal. He pledged that Canada would devote no less than 5 per cent of its R&D investment to a knowledge-based approach to the challenges faced by developing countries. The Martin government's new national science advisor was also tasked with holding discussions with the research community in Canada to identify further steps in development assistance, partnerships and capacity-building (Abdelgafar and Thorsteinsdottir 2007).

Under the Harper Conservatives, from 2006 to 2011, bio-policy seemed in some ways to be sublimated in its minority-government agenda – an agenda that focused initially on accountability reforms and on industrial sectors such as energy and forestry that were closer to its Western Canadian political base. A bit later the Harper government abolished the Canadian Biotechnology Advisory Committee as well as the Canadian Biotechnology Secretariat housed in Industry Canada, largely as a kind of efficiency-oriented part of its S&T strategy where a new overarching S&T advisory body was put in place.

The Harper government did not change or eliminate any biotech regulatory policies, and it did complete the planned establishment of Assisted Human Reproduction Canada during 2006. It was also reluctant to draw attention to the biotech policy field in its traditional food areas, even though in the bio-food realm in Saskatchewan and in bio-sciences in Alberta there was considerable support in the Conservatives' Western Canada prairie base of its core political and voter support. Instead, the Conservatives drew attention to their larger and more visible biofuels programs early in their mandate as a combined agricultural, energy, and environment initiative.

In 2009, however, the Harper government reduced and ended some of the funding for Genome Canada. Such actions raised suspicion among many observers that the former Reform Party elements of the governing Conservative party, including Prime Minister Harper and his science minister, were opposed to some of the bio-health and bio-life aspects of biotechnology research, partly on ethical and related religious grounds. But there was certainly no wholesale opposition when renewed support for Genome Canada was announced in the 2010 federal budget. In addition, the Conservatives embellished the website Canada Bio-Portal, which referred to a bio-strategy and the bio-economy, but mainly as a way for interested parties to get one-stop information about federal biotechnology policy and biotechnology in Canada overall.

So the Harper government's relative inattention or studied neutrality regarding bio-policy can just as easily be attributed to the government's simple preference to pay attention to other policy fields and funding priorities, including other industries such as oil and gas, forestry, and natural resources overall, and eventually the auto industry, once the full-blown 2008–9 recession hit the economies of Canada and other nations.

A crucial Harper-era biotech-related policy did emerge when the federal government released the long-awaited 2006 Blueprint for Renewal to transform Canada's approach to regulating health products and food, including a life-cycle approach and the practice of progressive licensing (Health Canada 2006a). It was long awaited, in that the Liberal government had earlier been working on similar ideas and kinds of policy and regulatory governance change. The Blueprint policy looms large in our account in chapter 5 of the bio-health realm. Meanwhile, other federal biotech policies were emerging or being debated in more specialized and even arm's-length arenas – some under CBAC advisory and advocacy auspices in the bio-health and bio-life realms, without garnering focused national political, partisan, or media attention.

Interacting with these national developments were closely related further international biotech support initiatives, such as the Human Genome Project and the mediating role of international science and networks of scientists engaged in risk-benefit choices and advice (Adams 1995; Smith and Halliwell 1999).

The political fault lines for biotechnology since 2000 divide the bio-food and bio-health realms, with extensions to bio-life realms. On the bio-health side are several major elements of state support for R&D, especially on initiatives such as the Human Genome Project. *Genomics* is defined as 'a discipline that aims to decipher and understand the entire genetic information content of an organism' (Genome Canada 1999, 7). It differs from classical genetic research in its large scale, broad scope, and heavy reliance on computer-based bio-informatics. There is little doubt, as later chapters show, that many commentators and organizations see it as the core of a new economy technology (Shreeve 2004; Thurow 1997; Wellcome Trust 1999).

The Human Genome Project is a global publicly funded project whose purpose is to map the human genome or the genetic 'book of humanity.' The scale of funding on the human genome project and on other genomics research has been impressive and is led largely by the United States and the United Kingdom, the latter partly through the Wellcome Trust. The U.S. National Institutes of Health helped to fund three genome centres, and overall American public funding easily exceeded $1 billion. Canada joined the genome project, and in the late 1990s, Germany, France, and Japan also committed large increases in genome funding through national government budgets.

Human genome research also proceeded at breakneck speed in the private sector, with the support of private equity alongside public investments. Understandably, this trend has resulted in a growing concern about ownership of genomic data. Some of these issues emerged visibly in 2000 when negotiations broke down between the Human Genome Project, the publicly funded multinational group, and Celera Genomics, a private U.S. company that was moving even faster than the public project to map the human genome. Other private biotechnology companies feared that Celera would lock up rights under patent protection. Public interest scientists and advocates feared that it would stifle the free exchange of scientific data and information (Shreeve 2004).

These concerns were reflected in a joint statement by President Bill Clinton and Prime Minister Tony Blair on 14 March 2000. The two leaders stressed the need for the human genome raw and fundamental data to be kept in the public domain so that scientists could utilize it freely. The Clinton-Blair statement highlighted their view that 'unencumbered access to this information will promote discoveries that will reduce the burden of disease, improve health around the world and enhance the quality of life for all humankind' (*Independent* 2000). Reflecting the complex interplay of values, interests, and political power in bio-policy, these leaders went on say that 'intellectual property protection for gene-based inventions will also play an important role in stimulating the development of important new health care products' (ibid.).

In effect, such initial rushed leadership statements are a form of moral suasion or exhortation that enter the international bio-policy regime norms, though not necessarily with a clear or unambiguous meaning; the short-term intended effect appears to be a blend of reassurances – to an anxious general public and concerned scientific community on one hand, to an ambitious and cheered market sector on the other. The speech in fact referred to both free and patented use, without specifying which was which. The exact nature of international bio-policy and related patenting thus remains contested both in market and public policy terms, and it has major implications as well for understanding both power and democracy in biotech governance.

The Core Politics Underlying International Biotech Policy

A final feature of this account of biotech policy history is how the core politics underlying international biotech policy have broadly changed over the last three decades. American power and interests, particularly in relation to the EU, are of particular importance, but U.S. direct influences also have a special continuing salience for Canada in a very intense cross-border and North American socio-political context.

The United States has been the strongest advocate of most of the trade-related aspects of the regime and of the push to ensure that biotechnology was not singled out as a new technology but rather was governed by generalized trade rules. U.S. political power was crucial in securing the TRIPs agreement and ensuring that intellectual property matters were put in the safer hands of the WTO rather than in broader bodies such as the World Intellectual Property Organization. Behind this concerted view was the massive lobbying power of the U.S. global pharmaceutical and biotechnology industry, and the supportive lobby within the U.S. went beyond these sectors as well, since other aspects of IP and trade were also involved in fields such as telecommunications.

In the basic thrust of the trade-related and TRIPS aspects of the regime, the European Union broadly aligned with the United States but was much more sceptical of biotechnology in food, because this was seen as a part of a U.S. and broader North American trade advantage and because public opinion in Europe strongly questioned the safety of biotechnology in food. As the 1990s ended, the fierce debate on biotechnology in the United Kingdom affected the climate of receptivity for biotechnology in Europe. The controversy over 'mad cow' disease also directly influenced this climate of debate in the EU and helped generate support for the precautionary approach and for a major reform of European food policy overall.

Biotechnology in food took on larger political importance in Europe. It could be linked in major countries such as France to aggressive pressure from the United States and elsewhere to open up European agriculture markets and to reduce European Union subsidies for agriculture. It also related to broader left-right politics and to the vibrant politics of new social movements in European countries where biotechnology evoked debates about the power of multinationals and about perceived unaccountable global governance by the international trade fraternity of policymakers (Hunt 1999; Lynas 1999).

EU biotechnology policy centred initially, therefore, on strong opposition to GM foods in particular and took the form of strong requirements for mandatory labelling of such foods and also outright refusals to approve many bio-food products. In part, this stance came about as well because few if any firms in the European Union were developing such products. This situation changed somewhat after the European Union was greatly enlarged to its current twenty-seven member states, with new member countries and some older member countries such as Germany beginning to seek approval for bio-food products.

In the bio-health and bio-life realms, the balance of power and interests within the European Union shifted in a major way. Here the politics were more positive overall, and numerous firms and universities and hospitals across member countries were engaged in research, seeking regulatory approvals through the European Medicines Agency.

The position of developing countries in the various international bio-policy processes is also noteworthy. Developing countries were typically defensive and comparatively weak in their political bargaining capacity. Broad trade-related measures ostensibly not specific to biotechnology were especially negotiated from a position of political weakness. Many developing countries signed on to the WTO because other trade-offs were made that were of interest to them. Moreover, they always faced the need to insert transition mechanisms to deal with their limitations in underlying scientific and technical capacity to implement trade and TRIPS provisions (Doern 1999; Lynas 1999; Trebilcock and Howse 1995). Developing countries thus continue to have strong concerns about the impact of biotechnology on biodiversity and on the ownership of the technology and its products.

The ascendancy of trade-related aspects of the emerging biotechnology regime arises from the fact that they were constructed and negotiated largely in the late 1980s and early 1990s. The latter two elements of the biotechnology regime traced above have emerged more in the late 1990s and into the first decade of the twenty-first century. Accordingly, the bio-safety protocol and even the Human Genome Project reflect a later set of political forces that broaden and deepen the notion of what a bio-policy regime might consist of or become, within a previously established, though still-forming regime of trade rules governing bio-food products and processes. Of course a larger of array of non-trade systemic power relations are involved nationally, in North America and globally, regarding developed and developing nations, drug companies and agri-food industries, as well as consumer interests and coalitions.

The bio-safety protocol is indeed overtly and directly focused upon parts of biotechnology. Of importance here is that by 1999–2000 public opinion in the United States was also changing toward biotechnology. Polling data showed, as in Europe, the United Kingdom, and Canada, that although there was support for biotechnology in health and in the human genome developments, concern was growing among Americans about GM food and consumer choice. This latter scepticism had not been present in earlier U.S. polls, since GM food was seen as a U.S. trade and industrial advantage, and American farmers supported GM foods.

Meanwhile, governments had relatively little difficulty justifying their support for the political 'high road' cause of human genome research, which could be positively linked to possible future cures for cancer and other diseases. At the same time, the human genome realm did raise the conundrum, and no simple riddle at that, of who owns the human genome data; hence, the initial Clinton-Blair exhortation to do the right thing without clearly defining the right thing in legally credible ways. At the end of the first decade of the twenty-first century, this balance has still not been agreed to in bio-policy terms.

For Canada, U.S. economic, regulatory, policy, and governance impacts have been important in ways that will emerge in more detail in the three bio-realm chapters. U.S. biotechnology policy on bio-food had direct cross-border impacts in part because of the lobbying of American agribusiness firms in Canada and because of the similar structure of large farms where bio-food could be tested, grown, and marketed. The intellectual property systems of both countries are tightly interwoven and affected by the same massive increases in patent application volumes during the formation of the bio-health realm. Frequently, as in recent years, the United States had tougher laws and rules, such as those regarding protection from discrimination due to genetic testing.

Other U.S. influences emerge from the tight day-to-day cooperation between Canadian and U.S. regulators such as Health Canada and the Food and Drug Administration. For example, there are direct links between Canada's Health Canada blueprint reforms in 2006 and 2007 and the major proposed amendments to the FDA's legislation in 2007. The latter involved strong shifts to post-market assessment, with important implications for the evidentiary and assessment criteria in the post-market compared to the pre-market (Evans 2009).

Conclusions

Canadian biotechnology policy is a set of linked and evolving national and international policies. This chapter has surveyed historically the major policy statements, strategies, and changes in the last two decades of the twentieth century and the first decade of the twenty-first century.

The main periods of Liberal and Conservative governments since 1980 show both partisan continuity and differences as biotech policy evolved nationally and internationally. The Trudeau Liberals adopted an initial strategy that overall was pro-food biotech. The Mulroney Conservatives and the later Chrétien Liberals broadened biotech policy to achieve a greater balance of values regarding environmental, health, and stewardship overall.

The Mulroney Conservatives, however, were the driving instigators of Canada's free trade agenda and influenced biotechnology through its adherence to trade-related norms, including those linked to sound science and stronger patent laws and intellectual property protection. The later Chrétien and Martin Liberal governments became enthusiastic supporters of genome-related research, mainly via new research funding bodies and network formation. The Liberals were more sluggish in some of the regulatory policy and governance aspects, including those regarding assisted human reproduction and related bio-health and bio-life aspects.

Finally and currently, the Harper Conservative government shows benign supports for biotech overall but also some concerns about Genome Canada funding and abolished the Canadian Biotech Advisory Committee. On the regulatory front, the Harper Conservative's blueprint strategy was advocating and promising a potentially stronger, more complex regime for post-market and also life-cycle approaches.

This historical picture overall does not suggest excessively partisan views and governing party differences. Mild partisanship may arise because, receiving only passing mentions in Speeches from the Throne and Budget Speeches, biotech policy operated in the middle and even lower-ranked realms of public policy and administration. Moreover, as the analysis of the three bio-realms will show, the actual underlying bio-technologies in bio-food, bio-health, and bio-life were not always occurring in orderly chronology.

The chapter shows the growing presence and complexity of multi-level governance as diverse national and international policies emerged and were implemented in biotechnology directly and indirectly in other policy areas that effect biotechnology such as patent, trade, environmental, health, and research policy. We also offered some initial basic observations about the core politics of global biotech policy, regarding both the European Union versus the United States and the continuous impact of U.S.-Canada relations.

Biotechnology policy cannot be divorced from underlying power relations, struggles, and alliances between nation states and powerful national and global business and other interests campaigning for, and often achieving, favourable rules and regimes. These power relations involve divisions between Europe and the United States in some bio-policy areas and considerable agreement in others. They also involve changing Canada-U.S. relations in trade, patent, and agriculture and health policies, as well as in the design of science, technology, and innovation policies. The nature of these power relationships is not a stable one, nor are the ideas underpinning them. Contested ideas on science and policy are subject to change in democratic debates, shaped by the shifting structure and power of increasingly networked interests.

PART TWO

Empirical Analysis of the Changing Biotech Governance Regime

3 Science and Supportive Governance

Introduction

We begin the detailed empirical examination of biotech governance with an analytical portrait of the science and support/promotion-related agencies in the Canadian biotech governance regime. Our enquiry in this chapter captures the nature and evolution of support for bio-science in five federal research agencies: the National Research Council (NRC) and its bio institutes and programs; the granting councils' jointly administered Networks of Centres of Excellence (NCE); the Canadian Institutes of Health Research (CIHR); the Canada Foundation for Innovation (CFI); and Genome Canada. In total, these agencies have elaborate networked links with businesses, universities and hospitals, mainly in Canada though extensively at an international level as well.

As in our discussion of biotech policy evolution, where we differentiated direct effects and indirect influences, this distinction applies too in the discussion of science and research agencies. Of the agencies examined in the first section, only one, Genome Canada, has a biotechnology focus. The other four agencies have broader research or research infrastructure mandates within which biotechnology is only one indirect dimension (Doern and Stoney 2009).

We then look, in the second section, at bio-policy and advisory bodies as arenas of promotion, debate, and consultation. These are cast as support/promotion-related in that they are undoubtedly part of the support governance structure for biotechnology. At the same time, however, these policy and advisory bodies deal with and partially accept the values and interests critical of, and/or opposed to, biotechnology. These public policy values and organized interests include: Industry Canada and its innovation and mainly pro-patenting agenda; the engagement of ministers in the interdepartmental bio-policy machinery anchored by the Canadian Biotechnology Secretariat (CBSEC); the external advisory and citizen engagement role of the Canadian Biotechnology Advisory Committee (CBAC); and periodic parliamentary committee inquiries, criticism, and involvement.

We also survey federal Budget Speeches and Throne Speeches from 1980 to 2010 to assess the policy agenda status of biotechnology. Taken as a whole, the analysis provides initial evidence for our argument that fundamental political realities explain why the Canadian biotechnology governance regime does not have a central point of authority within the state and why it has never been a high political priority but has instead functioned in the middle realms of agenda-setting.

The third section briefly examines longer-term and recent efforts in Canada to organize more formal kinds of technology assessment as a feature of democratic government. While these efforts go beyond bio-technology, the difficulty in devising such policy and deliberative processes is germane to our assessment of the democratic strengths and weaknesses of Canada's biotechnology governance regime, including its science-based and precautionary elements.

Federal and Related Biotech Research Agencies: Networked Governance, Support, and Research Conduct

The changing complexity of federal government support for bio-research of various strands has occurred mainly through research agencies and research funding entities that represent an increasingly networked form of governance, support, and the actual conduct of biotechnology research. We discuss five such bodies here. Among federal granting councils, we look especially at the CIHR for reasons explained below. Some reference is also made to the two other granting councils, NSERC and SSHRC, which fund some biotech-related research under their normal grant programs where individual researchers apply for typically smaller grants.

Therefore, rather than provide a detailed account of substantive bio-research in various agencies, we have chosen to present the overall patterns of agency support and governance on the research side. We look briefly in turn at each agency in the context of each agency's own basic

Table 3.1: Federal bio-science and research support entities

National Research Council

- Institute for Marine Biosciences in Halifax
- Biotechnology Research Institute in Montreal
- Institute for Biological Science in Ottawa
- Institute for Biodiagnostics in Winnipeg
- Plant Biotechnology Institute in Saskatoon
- Genomics and Health Initiative in Ottawa

The Granting Councils' Networks of Centres of Excellence

- Insect Biotech Canada network (established in 1989)
- Stem Cell Network (2008)
- Centre of Excellence in Personalized Medicine (2009)

Canada Foundation for Innovation

- 709 biotechnology infrastructure projects funded since 1998 mainly in bio-health and bio-life
- Over $766 million of CFI funds invested, as well as required partner funding that amounts to a further approximately $1.5 billion, with partner funding coming mainly from the provinces

Canadian Institutes of Health Research

- Institute of Genetics
- All of the CIHR's thirteen institutes are expected to conduct and relate their health research to the CIHR's four pillars of health sciences inquiry: bio-medical, clinical, health systems and services, and population health

Genome Canada

- Focus on genomics and proteomics research
- Six regional Genome Centres across Canada
- Mandate role in the ethical, environmental, economic, legal, and social issues (GE^{3}LS) associated with genomics and proteomics research

---

evolution in federal research and innovation and in network formation. Table 3.1 helps as a guide to this narrative.

National Research Council

Canada's best-known federal research institution, the NRC, has been supporting basic and applied research in numerous areas of research, especially during and since the Second World War. Following its post-war heyday, it was criticized by early science-policy critics as being too isolated and described often by such critics as functioning 'like a university without students.' The NRC was under increasing pressure in the 1980s and early 1990s to become more commercially oriented and focused upon innovation (Doern and Levesque 2002).

During the 1980s and 1990s, NRC established five bio-institutes, which were created from the outset as NRC institutes rather than divisions and hence carried far fewer discipline-based traditions of former NRC divisions in NRC's evolution. The institutes were deeply influenced by views about natural or existing strengths in regional and local innovation systems, and also in ideas about an industry greatly dependent on supportive federal policy, including policies on intellectual property.

The larger bio-focus of the NRC dates back to 1983 and includes five fields of specialization, the first two dealing with human applications (human diagnostics and human therapeutics) and the three remaining fields concerning biotechnology applications in agriculture, aquaculture, and environment. As the twenty-first century ensues, each of the five NRC institutes specializes in one of these fields of research: the Institute for Marine Biosciences in Halifax in aquaculture, bio-informatics, and sequencing; the Biotechnology Research Institute in Montreal in human therapeutics as well as the environment; the Institute for Biological Science in Ottawa in immune-chemistry; the Institute for Biodiagnostics in Winnipeg in human diagnostics; and finally, the Plant Biotechnology Institute in Saskatoon in agriculture.

The Biotechnology Research Institute (BRI) is a practical example of the five NRC bio-institutes. Its vision, as stated in 1999, is to 'create an internationally recognized incubator facility for the Canadian biotechnology industry,' and its stated mission is to 'contribute to wealth generation and job creation in the environmental biotechnology and pharmaceutical industries, to pursue world-class, innovative research, which through Institute licenses, generates products and processes for industry, and to foster the needed skills and knowledge to maintain Canadian leadership in the field' (National Research Council Canada 1999, 3).

BRI's R&D strategy focuses on two main areas of industrial activity: the pharmaceutical and natural resource (environment) industries. Its Pharmaceutical Biotechnology Sector applies biotechnologies to assist in the development of novel strategies for the treatment of cancer, and cardiovascular and inflammatory diseases. Its researchers identify new molecular targets and develop and improve potential therapeutic agents using drug design. BRI's Bioprocess Sector carries out bio-process innovation and technological development from process inception to industrial-level production, using microorganisms, enzymes, or animal cells as bio-catalysts. These bio-processes concentrate on two areas of key importance in environmental activity: the remediation of polluted sites, and the problems posed by the increase in atmospheric greenhouse gases.

BRI began its work in 1987, a direct product of the 1983 strategy on biotechnology. A significant part of that strategy was centred on Montreal's pharmaceutical industry and hence it was a reasonably obvious choice to locate the BRI in Montreal.

The other NRC bio-institutes have all developed in their local-regional contexts, with varying degrees of explicit commercial motivation, depending on the inherent nature of the bio-sciences and bio-research on which they were focusing. By the late 1990s, all the NRC bio-institutes described their roles as being anchored in clusters or linkages with other networked partners whose skills, knowledge, or financing were needed to bring products and technologies into applied use. Of the five, other than the BRI, the Plant Biotechnology Institute (PBI) centred in Saskatoon is the most dynamic and best known internationally. It is fully networked with the key bio-food and crop firms, such as Dow AgroSciences Canada in canola research, and with the University of Saskatchewan and the government of Saskatchewan (National Research Council Canada 2010b).

In the 1990s, NRC also started the Genomics and Health Initiative, which, while not an institute per se, gives more explicit focus to genomics. Centred in Ottawa, its work is to advance genome and health research in areas including infectious disease detection and treatment, bio-renewable oils, cancer therapeutics, heart disease management, and virtual reality systems for surgical oncology (National Research Council 2010a).

The Granting Councils' Networks of Centres of Excellence

Bio-research is also supported through federal funding and network-based entities such as Networks of Centres of Excellence. Managed jointly by the three federal granting councils – NSERC, SSHRC, and CHIR – the NCE began in 1987 with the explicit purpose of fostering networks of researchers across Canada in different current and emerging research realms (Atkinson-Grosjean 2006). Along with the changes noted above regarding the NRC, the network focus has sought to place greater emphasis on interdisciplinary research across the natural, health, and social sciences.

NCE grants, awarded every few years in new rounds of application and competition, brought more visibly to the surface *collectivities* of researchers, in geographically dispersed virtual reality networks, but in practice also identified with certain regions or clusters of big versus small universities (Atkinson-Grosjean 2006). At some level, this creates political concerns about accountability for the adequacy of the regional distribution of funds, which are increasingly valid if some kinds of financial support such as NCE funding require applicants to 'bring money to get money.' Levered money and/or partnership funding, while not dominant in the overall granting-body funding picture, is nonetheless a growing part of it in the last ten years, as well as in infrastructure funding provided by funding entities such as the Canada Foundation for Innovation.

Inevitably, the NCE networks created concerns among academics and scientists about the excessive commercialization of university research (Atkinson-Grosjean 2006; Doern and Stoney 2009). Despite the expression of such concerns, under the Harper Conservative government, commercialization goals became more overt and more strongly encouraged. Under the NCE apparatus two new programs were launched: the Centres of Excellence for Commercialization and Research, and Business-Led Networks of Centres of Excellence (Networks of Centres of Excellence 2010).

Bio-research networks emerged early in the NCE funding process. In 1989, the Insect Biotech Canada network was funded, centred at Queen's University and devoted to biotechnology for insect pest management. At present, NCE networks include two that are more focused on bio-health and bio-life: the Stem Cell Network and the Centre of Excellence in Personalized Medicine (Networks of Centres of Excellence 2010).

In the current grouping of its networks, NCE assigns biotechnology as an overall descriptor alongside health and human development, in which there are six named networks, including an Advanced Foods and Materials Network and an Allergy, Genes, and Environment Network. At its 2009 annual meeting, NCE held ten research and network reporting and discussion sessions grouped under Biotechnology, Life Sciences, and Pharmaceuticals (Networks of Centres of Excellence

2010). These groupings show that crossover boundaries of bio-research are increasingly the norm and, of course, are part of why NCE formed in the first place, not for bio-research by itself but for all forms of actual and potential networked and interdisciplinary research.

Canada Foundation for Innovation

The Canada Foundation for Innovation (CFI) was created by the Chrétien Liberal government in 1997 to support research infrastructure in universities and research hospitals (Canada Foundation for Innovation 2010; Lopreite and Murphy 2009). The federal government had not supported such university infrastructure before in any concerted way. Formed as an independent foundation, the CFI was given an initial budget of $800 million with large later increases coming in other federal budgets in the late 1990s and beyond. The Harper Conservative government has given it a further $510 million in investment funding.

A leading example of the foundation model in use by the federal government, the CFI has been criticized as an arm's-length organization dispensing taxpayers' money (its initial and only endowment) without being accountable in the 'normal' way to Parliament and elected politicians (Aucoin 2003). The CFI's requirement that all universities (and research hospitals) bidding for infrastructure funds must submit a strategic research plan changed decision processes within universities, few of which previously tended to have such strategic plans. The CFI also has rule-making elements in its mandate simply because its spending comes with rules requirements to bring money from other funding partners (public or private) in order to acquire CFI money.

The CFI's funding of biotechnology infrastructure and platforms has been significant, involving, as per its innovation mandate, an array of smaller and larger universities across Canada as well as hospitals and organizations such as cancer charities and organizations. By early 2010, 720 biotechnology projects were funded, amounting to over $766 million of CFI money. When paired with even larger amounts of the required co-funding, the totals reach a further $1.5 billion, which has tended to come mainly from the provinces but also from business (Canada Foundation for Innovation 2010). The size of bio-infrastructure projects extends from those in the $100,000 range of CFI money to several in the multi-million-dollar range, with projects mainly in the bio-health and bio-life realms.

Canada Institutes of Health Research

Among the three main federal research granting councils, we single out the CIHR for analysis, given its unambiguous health focus and because it went further than the other granting institutions (NSERC and SSHRC) in explicitly fostering applied interdisciplinary research networks within which bio-research had further room to flourish.

In 2000, the Chrétien Liberal government formed the CIHR by merging the former Medical Research Council of Canada and the National Health and Development Research Program from Health Canada and, in the process, by creating 'virtual reality' research institutes devoted to interdisciplinary research on applied health and wellness problems and challenges (Murphy 2007).

Its initial set of thirteen institutes emerged from further consultation and includes institutes on traditional or familiar areas of research, such as one on cancer research, and also several that meet the test of broader interdisciplinary realms, such as the bio-research-focused Institute of Genetics (see more below). The thirteen institutes include groupings for:

- Aboriginal people's health
- Aging
- Cancer research
- Circulatory and respiratory health
- Gender and health
- Genetics
- Health services and policy research
- Human development, child and youth health
- Infection and immunology
- Musculoskeletal health and arthritis
- Neurosciences, mental health, and addiction
- Nutrition, metabolism, and diabetes
- Population and public health (Canada 2005, 4)

Federal policy in 2000 stressed that 'the Institutes would not be centralized "bricks and mortar" facilities. Instead these virtual organizations would support and link researchers who may be located in universities, hospitals and other research centres across Canada' (ibid.). Another key feature of the institutes is that they 'would support researchers who approach health challenges from different disciplinary perspectives' (ibid.). The vision of transformation that the CIHR was to bring saw the old system and model as having '*dispersed* research efforts, disciplinary *separation, separate* from delivery, and *multiple agendas*,' whereas the new model anchored in the CIHR would be '*integrated*: across geography/ institutions; across scientific disciplines; into the health system; and with the national health agenda' (ibid., 6; emphasis added).

Each of these institutes also expects to conduct research across the four pillars of health sciences inquiry: bio-medical, clinical, health systems and services, and population health. The notion of bio-medical research as one of the four pillars means that all of the institutes have some expected link to bio-medical research.

Among the three federal granting councils, the CIHR has garnered by far the largest percentage increases in spending in the last decade, three times that of the NSERC and the SSHRC (Doern 2009; Murphy 2007). In part, this difference is due to the greater political saliency of health issues among voters compared to science and engineering or the social sciences. The CIHR also has a commercialization focus, partly because its emergence overlapped with the federal innovation strategy in 2002 and the greater expectation of research that would support a health 'industry' and not just a health sector. These commercialization goals, however, have been a difficult challenge, given that the large majority of the health research community sees itself as being in a public interest–centred research endeavour overall, rather than a commercial one.

The most obvious institutional presence of bio-research in the CIHR is the above mentioned Institute of Genetics. Its mandate is to support research 'on the human and model genomes and on all aspects of genetics, basic biochemistry and cell biology related to health and disease, including the translation of knowledge into health policy and practice, and the societal implications of genetic discoveries' (Institute of Genetics 2009). In keeping with this mandate, among its research priority themes are: from genes to genomic medicine; population genetics, genetic epidemiology, and complex diseases; and genetics and ethical, legal, and social issues.

Then again, the bio-research presence is not found just in the IG. It also shows up in the structure of the CIHR's forty-seven committees that review applications and peer review. The peer review committees have quadrupled since the days of the former Medical Research Council that the CIHR replaced. At least six of these review committees are oriented to bio-research.

Beside the CIHR, the other two granting councils at the national level fund biotechnology research in their normal grants to individual and small teams of academics across the rest of the natural and social sciences. For example, SSHRC funding of biotechnology-related social science research involved 125 projects from 1998–9 to 2008–9 (Social Sciences and Humanities Research Council of Canada 2010). Not surprisingly, a significant portion of SSHRC grants involved research on social and ethical concerns about biotechnology.

Granting data from NSERC are not as systematically available or categorized as clearly, but in 2007–8 NSERC shows that collaborative R&D grants expenditures in the 'food and bio-industries' sector amounted to about $10 million of NSERC funding and $12 million by industry (Natural Science and Engineering Research Council 2008).

Genome Canada

Genome Canada, the final federal research agency we survey, is the only one of the four that is bio-focused in its central mandate. Formed initially through a funding investment of $160 million in the 2000 federal budget, Genome Canada was established as a foundation similar to the Canada Foundation for Innovation and other arm's-length foundations being formed during this period (KPMG 2007). The name Genome Canada had actually been the title of a group formed in the late 1990s composed of a network of scientists from hospitals, research institutes, and the public sector (Gelineault 2002). During this same period, other informal groups in different regions of Canada, and in bodies such as the MRC and also in the National Biotechnology Advisory Committee (later changed to CBAC), had lobbied the federal government with their concerns that Canada was not keeping up with funding and support in this crucial field of research in the life sciences. As we noted previously, 2000 was also the year in which U.S. President Bill Clinton and U.K. Prime Minister Tony Blair announced that the human genome had been mapped.

Genome Canada's mandate is 'to develop and implement a national strategy for supporting large-scale genomics and proteomics research projects for the benefit of all Canadians' (Genome Canada 2009b). Its emphasis is on the delivery of 'tangible and measurable results ... in research in human health, agriculture, environment, forestry, fisheries, and new technology development' (Genome Canada 2009b). Genome Canada declares that it seeks to 'play a leadership role on the ethical, environmental, economic, legal and social issues (GE³LS) associated with genomics and proteomics research' (Genome Canada 2009a). We have more to say about the GE³LS aspect of its mandate in the second section of this chapter.

Since 2000, the federal government has invested $840 million in Genome Canada, which has been levered into almost $1 billion in partnered co-funding and interest earnings. Funding partners have included public, private, and venture philanthropist organizations both in Canada and internationally. Genome Canada provides up to 50 per cent of the funding for large-scale research projects, or 100 per cent for science and technology platforms that provide access for researchers to sophisticated technology and expensive research infrastructure (KPMG 2009, 3). In this latter platform and infrastructure role, it has some similarity with the larger infrastructure mandate of the CFI. By late 2008, Genome Canada had funded over 100 large-scale research projects and ten science and technology platforms (ibid.).

Genome Canada has also established six regional Genome Centres across Canada. These regional centres provide different focal points for research and possible commercialization, but because the regions vary in their own ability to raise the required partnered/networked funding, there are significant built-in regional disparities as well. The Ontario Genomics Institute, Genome British Columbia, and Genome Alberta have the best access to partnered funding, including provincial government funding together with some international funding.

Genome Canada's guidelines and evaluation criteria for these centres state that they must conduct research into the above-noted environmental, ethical, legal, and social issues related to genomics (Einseidel and Timmermans 2005; Williams 2006). From the outset, apprehensions have been expressed that the GE³LS mandate concerns about community building and wider debate are not being given sufficient attention and priority (Gelineault 2002). The seeming marginalization of the GE³LS mandate regarding these questions that lie at the interface between science and society, was further reflected in 2009 when Genome Canada's website provided for the first time a GE³LS webpage 'now directly accessible from our home page' so as 'to emphasize GE³LS right up front' and show 'how integral GE³LS research is to the advancement and application of genomics science in Canada' (Genome Canada 2009a). Genome Quebec (2009) seems to have the most active and wide-ranging GE³LS, including purposeful discussion and debate on the commercialization issues regarding genomic research in Canada (Caulfield 2009b; Cook-Deegan 2009; Landry 2009).

An evaluation of Genome Canada by KPMG published in 2009 was exceedingly positive, concluding that it has had 'a tremendous impact on genomics and related research in Canada and its rationale remains strong' and that 'it has transformed the quality and quantity' of genomics research, and furthermore has been 'important for attraction and (especially) retention of faculty members' (KPMG 2009, 3–4).

Forged in the Chrétien Liberal government era and backed in the brief Martin Liberal government period from 2004 to 2006, Genome Canada initially received further support in 2006–7 by the Harper Conservative government. Yet, in the 2009 federal budget, Genome Canada lost out in the first recession-era budget and did not gain its usual new cash injection (Abraham 2009). As well, other cuts occurred to the three main granting councils, the net effect of which was a growing concern among scientists that Genome Canada support was in particular being targeted for cuts. However, in the 2010 federal budget, $75 million in Genome Canada funding was restored to previous levels, in part in reaction to the criticisms.

Biotech Policy and Advisory Bodies and Arenas of Partial Support, Debate, Criticism, and Consultation

If the mainly bio-supportive role of the research institutions is rather apparent, the entities and arenas of support, debate, criticism, and consultation are more nuanced. We look in turn at Industry Canada and its innovation and related intellectual property agenda; the inter-departmental bio-policy machinery, including the low profile that biotechnology has had in Speeches from the Throne and Budget Speeches; the Canadian Biotechnology Advisory Committee; and periodic parliamentary and opposition political party involvement, including committee inquiries that focused on biotechnology.

Our analysis shows in concrete terms the nature of the political realities that ensured that biotechnology was rarely even mentioned at the top of the policy and political agenda; instead, biotechnology has functioned at middle levels and often, even there, in quite indirect ways, buried or subsumed in discussion and discourse on innovation, health research, and broader umbrella concepts and ways of expressing national priorities.

Industry Canada and Its Innovation and
Related Intellectual Property Agenda

Industry Canada, as the lead industrial, science, and technology and innovation policy department of the federal government, has openly been a source of overall support for both bio-research and for the bio-technology industry. For much of the period covered by this book, Industry Canada (and its immediate predecessor differently named 'industry' departments) functioned under the broad policy rubric of innovation policy. This policy role includes broad support for greater patenting by Canadian companies (biotechnology and others) and also by universities as a contributing engine and indicator of innovation (Doern and Sharaput 2000). Industry Canada saw the national and global property-protection aspects of patents as crucial and did not particularly acknowledge the criticisms of the role of patents that focused more on knowledge dissemination that were expressed then and emerged more strongly from 2000 on (Castle 2009; Caulfield 2009a; International Expert Group on Biotechnology, Innovation and Intellectual Property 2008; May 2009).

In both the core innovation policy realm and in the bio-science and bio-industry aspects, Industry Canada knew in reality that it was not the only federal department involved in these policy and governance realms (Doern 1996; Doern and Kinder 2007). Moreover, biotechnology was not the only industrial sector or enabling technology it was interested in and concerned about. It had to deal with several industries of both a traditional sectoral kind (for example, automobile, cultural) and the enabling kind (for example, Internet and nanotechnology – both with growing links with biotechnology).

Industry Canada also had major mandate regulatory tasks regarding marketplace governance in realms such as competition, intellectual property, corporations, and consumer policy. Many of these areas had potential impacts, positive and negative, on all three bio-realms as they came on stream, indicating the interplay of bio-policy with other areas in industrial policy.

In addition, located physically within it until their demise were both the CBSEC, which anchored the interdepartmental committee of officials involved in biotech policy and management, and CBAC, which functioned as an arm's-length body, but whose secretariat was CBSEC. In short, there was not much of a firewall there but nonetheless,

Industry Canada had to be subtle and sophisticated about its support for bio-industries and in the way it debated biotechnology within the Cabinet and federal bureaucracy, and tried to advance bio-policy and bio-governance change.

For the 1970s and 1980s, the early part of the period we are covering, Industry Canada's information and discourse focused on biotechnology in agriculture or bio-food. But this supportive role was shared with and also dominated by Agriculture Canada (later and presently Agriculture and Agri-Food Canada), renamed precisely to give greater emphasis to the secondary but growing food sector as an industry, as opposed to the previous focus on primary agricultural producers/farmers (Doern 2004; Prince 2000; Skogstad 2008). While Industry Canada clearly saw bio-food as a strategic and growing sector, its officials and ministers, given other industry priorities, were never keen about putting their heads above the political barricades for bio-food, given some of the latter's Frankenfood discourse and outright unpopularity for most ministers.

All this changed in the 1990s and 2000s, when bio-health and bio-life realms emerged more clearly. Industry Canada relocated its biotechnology focus into its life-sciences sector (and its website life-sciences gateway). It stresses there that 'Canada's biotechnology sector has expanded rapidly in the last decade, in terms of industry-wide revenues, the launch of new companies, and in the continued diversification of products. Canada is one of the top five countries in the world in this vital field' (Industry Canada 2009a).

Industry Canada draws particular attention to the pharmaceuticals and biopharmaceuticals sector of the life-sciences industry. Some of this shift in focus was the product of an early Harper-era report by Industry Canada that examined public engagement on the future federal role in biotechnology (Decima Research 2006). Industry Canada also issued a progress report on the Harper government's 2007 report on *Mobilizing Science and Technology to Canada's Advantage* (Industry Canada 2007). The progress report on the early work of its newly created Science, Technology, and Innovation Council (STIC) drew attention to STIC's recommended set of four sub-priorities for S&T support. One of the four is 'health and related life sciences and technologies,' within which is listed biomedical engineering and medical technologies (Industry Canada 2009b).

Again, bio-life becomes the larger rubric for discussion while biotechnology overall does not leap out for special mention. Industry

Canada itself, however, does draw attention to the now far larger set of 'clusters of biotechnology activity' across the country.

It is mainly bio-health expansion that makes this new visible mapping and discourse possible, and Industry Canada shows such clusters and networks of various sizes and complexity in twenty-one cities and regions across Canada.

Engaging Ministers in the Interdepartmental
Bio-Policy and Governance Structure

Regarding ministerial involvement, we look first at where biotechnology stood in agenda-setting and profile among prime ministers and ministers of finance, as revealed through Speeches from the Throne (SFT) and Budget Speeches. Table 3.2 provides a summary of the biotech presence in this crucial level of political profiling. From 1980 to 2010, biotechnology is mentioned in seven of the twenty SFTs and then in only one or two words.

In the later years of the Trudeau era, biotechnology was never mentioned as a national priority in any SFT. In the Mulroney era, biotechnology was mentioned in the 1989 SFT as an example of a strategic technology. In the Chrétien and Martin Liberal era from 1994 to 2005, biotechnology is mentioned in four SFTs: in 1996 as an enabling technology; in 1997 as bio-pharmaceuticals and biotech in agriculture and fisheries; in 2001 as 'life sciences'; and in the two SFTs of 2004 as an enabling technology. Innovation as a national priority received top attention in the 2001 Throne Speech following Chrétien's late 2000 third mandate election victory, but even here, biotechnology receives mention only briefly as life sciences (Doern 2002a; Hale 2002). In the Harper era since 2006, biotechnology gained mention as 'bio-fuels' in the 2007 SFT as a part of supporting Canada's traditional natural resource industries. In the other five Harper-era SFTs to date, biotech is not mentioned at all.[1]

Budget Speeches and plans delivered regularly and publicly are the central occasion in which a government expresses policy choices and communicates resource/spending decisions to the general public and to political interests. As with Throne Speeches, biotechnology was rarely given an explicit highlighted mention in Budget Speeches. In thirty-four federal Budget Speeches from 1980 to 2010, a biotech mention occurs in only six.[2] There were no references to biotech in either the Trudeau or Mulroney eras. During the Chrétien and Martin era from 1994 to

Table 3.2: Biotechnology as a low-profile presence in federal Throne Speeches and Budget Speeches, 1980–2010

Federal regimes	Throne Speeches (n = 20) references	Budget Speeches (n = 34) references
Trudeau (1980–4)	0	0
Mulroney (1984–93)	1	0
Chrétien-Martin (1994–2005)	5	5
Harper (2006–10)	1	1
Totals	7 (35%)	6 (17%)

2005, biotech is mentioned in five Budget Speeches, mainly regarding Genome Canada and as a medical research biotech issue, especially when Paul Martin was the minister of finance and aggressively pursuing an S&T and innovation agenda when federal budgetary surpluses emerged from 1997 onward. In the Harper-era Budget Speeches, biotech is mentioned in one Budget Speech in 2007 with announced funding for the Life Sciences Research Institute in Nova Scotia. Even when the restoration of Genome Canada funding occurred in the 2010 budget, outwardly a good news item, it was not mentioned in the actual Budget Speech.

SFTs and Budget Speeches are especially indicative of what the prime minister and minister of finance are prepared to draw attention to in their own ultimate narrative of the priorities and agenda themes of the government. In this regard, of the thirteen total mentions in the fifty-four total speech documents, ten were during the Chrétien-Martin years, but even here they were subsumed mainly under their broader innovation, S&T, and health policy agenda-setting and thematic narratives.

Regarding both sets of major speeches, biotechnology is mentioned mainly in middle- and lower-level policy statements rather than in major agenda-setting and agenda-announcing occasions. This thirty-year account also shows the types of political language used in these agenda speeches. One is generic or high level, calling biotech strategic or enabling, thus tapping into dominant societal beliefs about science and technology. The second and more common type is discourse specific to one or other realm or industrial sector, such as life science, medical research, agriculture, bio-fuels, or pharmaceuticals. Both types of political language are designed to support and promote biotechnology through either the use of an abstract yet positively evocative word, such as *enabling*, or of an identifiable economic application or research entity.

For most of the last decade, the interdepartmental bio-policy and governance structure within the federal government has consisted of a committee of Cabinet ministers, and the Biotechnology Ministerial Coordinating Committee, with biotechnology mandates and a supporting committee of assistant deputy ministers. Both committees met infrequently. Both were served by CBSEC and its small staff. CBSEC and the ADM-level committee had a modest budget, including a fund for projects or initiatives submitted by biotechnology departments/ agencies.

In 2003–4, a key potentially different contextual factor was the need to engage a new prime minister, Paul Martin, who was succeeding Jean Chrétien. Martin, as we have seen, was a sympathetic leader on the subject of biotechnology and had been the central figure as minister of finance in the Chrétien years in garnering massive new research and innovation funds overall, including the agencies discussed above. With Martin's arrival as prime minister also came a changed set of ministers in the biotech mandate departments and agencies that had to be brought up to speed about biotechnology and its changing policy, and regulatory and governance challenges.

In particular, there was a need to develop a full appreciation by ministers and all stakeholders about what the broadened notion of the biotechnology regulatory regime now comprised. The shift to bio-health and the growing volume of products was bringing into the bio regime – alongside the traditionally included regulators such as Health Canada, the CFIA, and Environment Canada – the roles of patent agencies, formulary and pricing authorities under health care and pharmaceutical drugs, as well as research-ethics regulators.

In anticipation of a new Martin government, CBSEC commissioned a set of unpublished papers to review and debate the internal overall governance structure, including CBAC. The reviews were positive about CBAC's role, but all of them in their own ways emphasized that ministers had not been sufficiently engaged in the biotechnology policy file. There were suggestions in the resulting discussion that biotech on the food side had become a 'no-go' area for ministers because they tended to see it as mainly a political negative.

Some possible new mechanisms to engage ministers were then briefly discussed within the bureaucracy. One idea was establishing a Cabinet reference committee on biotechnology or the bio-economy.

Reference committees had been established just prior to this review period on energy policy (in North America) and on climate change. They were not decision-making committees but rather received a mandate from the prime minister to report back to him on the matter referred. This process allows ministers over a brief period to examine, study, and discuss the breadth of the issues. Thereafter the policy process would revert to normal Cabinet committee proposals and decision-making.

Another suggestion was, as a parliamentary complement to this Cabinet process, the establishment of a possible standing or temporary committee of the House of Commons or the Senate to examine the overall issues of enabling technologies, central to which would be biotechnology, linked to the growing convergence among Internet and information technologies and nanotechnology.

Neither of these structural changes garnered support in the federal public service bureaucracy, but the overall discussion did result in the draft blueprint for biotechnology (Canadian Biotechnology Secretariat 2004). It was informally agreed to by the committee of ADMs but never garnered ministerial approval. The blueprint suggested a framework that would position Canada as a world leader in biotechnology and its applications. Its overall objective was to accelerate the commercialization of biotechnology research for the social, environmental, and economic benefit of Canadians. The blueprint process centred also on getting ready for the larger bio-economy, integrating an economic agenda with effective stewardship and promoting a comprehensive systems approach with a global approach in federal decision-making.

Once again it was difficult to engage and hold ministers to a firmer and more visible and high-priority biotech policy and biotech governance commitment, even though the bio-health and bio-life aspects were in principle more positive politically. These themes of discourse were present in Industry Canada and in varied ways in the research agencies, but ministers in the main had other preoccupations and priorities both during this period and in the later Harper Conservative government years from 2006 onward.

A theme here is the difficulty of securing ministerial engagement across the government machinery. It must be remembered, however, that ministers were also being advised by their own department bio-advisors as distinct from CBSEC and the larger committee process. And, also crucially, Cabinet ministers were being lobbied directly by bio-firms, BIOTECanada, and numerous NGOs that were not particularly enamoured with structural change for its own sake.

Evidence in regards to the machinery of government question, as well as the relatively low priority and profile in Budget and Throne Speeches, suggests biotechnology functions in the relatively subdued middle worlds of political life and governance. Several possible reasons come to mind and warrant further brief discussion, especially when Canada is compared to the European Union and United States. Canada's biotech agenda can be contrasted in part with a higher-profile political presence at times in Europe, where bio-food opposition was strong and sustained, partly because of the relative absence of bio-food producers, compared with North America. There has also been a higher periodic presence in the United States, in part because of the stronger role of religion in U.S. politics and in bio-life issues. Further reasons that need to be taken into account, include (1) the reality that ministers and politicians often do not quite know how to deal with biotechnology in a raw political sense in any of its bio-food, bio-health, and bio-life realms/eras; (2) the fact that the bio-world involves complex kinds of science and technology that are frequently beyond the capacity of non-scientist politicians to discuss and communicate comfortably and with some clarity; and (3) precisely because bio-technologies and their governance deal with boundary-breaking notions of the public and private and the social and economic, they yield further discomfort for politicians who must try to explain complex forces and values in an otherwise sound-bite age of political discourse and criticism. In addition, there is the plain fact that few policy fields always make it to the top of the national agenda, since the agenda is crowded with competing demands, pressing time, and limited money.

The first suggested reason may have some validity. Bio-food and its frequent Frankenfood discourse have caused difficulties and led to embedded views that bio-food politics on balance is negative. But these views did not prevent support for bio-food in parts of the Cabinet and bureaucracy. And of course in the European Union, where bio-food was sharply criticized, that disapproval was also augmented by specific arguments about the need for information and choice for consumers and decisions to ensure that these needs were met.

The overall negatives seem not to be the case overall for bio-health and aspects of bio-life. Here the elemental politics are positive in the sense of searches for the causes of and cures for diseases compared to bio-food, but not uniformly so, especially in bio-life, where religious beliefs and discourse and concerns about the nature of life itself are not easy. However, as noted in chapter 2, these concerns were raised to a high national level of priority and concern in the United States.

The second science content-related reason seems to ring true on the whole. As the next three bio-realm chapters show, science-based assessments and approvals of particular products tend not to be raised by politicians, because they are not scientists and do not have the expertise. However occasionally, as in the rbST case (see more below), science disputes do get coverage in parliamentary committees, where more time is available to focus on them. Politicians are periodically interested in and capable of discussing some of the democratic implications about science and government in the various bio-realms.

The final public-private boundary-breaking suggestion as a reason is quite plausible in all three bio-realms. In bio-food issues of what is 'natural' food versus 'unnatural' products create difficulties as boundary issues, including public-private ones. The same is true in different ways for bio-health, where patenting volumes in health products created dilemmas about the commodification of research and the human body and of the boundaries of public goods and private property. And bio-life creates great discomfort and strong views about one's own body and its sanctity seen in many complex ways.

CBAC, External Advice, and Governance through
Citizen Engagement

CBAC as a non-statutory committee of twelve to twenty experts was established in 1999 with a mandate to 'provide comprehensive advice on current policy issues associated with ethical, legal, social, regulatory, economic, scientific, environmental, and health aspects of biotechnology' (Health Canada 2010a). In particular, it was to engage Canadians in the dialogue and to provide them with 'accurate information' about these biotechnology issues. CBAC members were nominated through a public process, with names reviewed by a Biotechnology Deputy Ministers Selection Panel according to specific criteria such as expertise, knowledge, and experience.

While CBAC was at arm's length, in some sense, from the government, it was frequently criticized, especially in its initial years, as being illegitimate and too close to the federal government's overall view of biotechnology in its whole approach to consultation. For example, a 2001 news release by a coalition of NGOs announced that it would not take part in CBAC's then initial consultations (Council of Canadians and Greenpeace 2001). This opposition abated somewhat in CBAC's later years, but it has always been to some degree a continuing feature of the larger governance and politics of bio-technology on its various bio-realms, especially regarding bio-food (Peekhaus 2010). The review process in 2003–4, noted above, assessed the role of CBAC, broadly concluding that CBAC did good work and addressed important bio-technology issues. From 1999 to its demise in 2007, CBAC published and discussed over seventy-five reports and studies, and specific items of advice to the government. These included reports on periodic national and regional roundtable discussions, as well as on subjects that included GM foods, the patenting of life forms, human genetic materials, an economic profile of the biotech sector, the precautionary principle, primordial stem-cell regulation, and population bio-banking. The studies showed a bio-food focus initially, but overall by far the largest number of studies and reports were in bio-health and bio-life realms. Without doubt, CBAC produced and published the main core of published policy and governance-relevant bio-analysis in Canada, but in 2010 its website was closed by the Harper Conservatives.

Despite this good record of research, Canadian Cabinets under the Liberals and Conservatives did not often actively seek advice from CBAC. CBAC therefore faced practical limits on the timeliness and use of its studies and advice. When feeding into a complex government-wide agenda, where there are many advisory bodies and many advisors, no single advisory body can expect its advice to be always accepted or always listened to, or even read. This was certainly true of earlier bodies such as the Economic Council of Canada and the Science Council of Canada, and it was true for CBAC as well. CBAC was by no means the only arena for such involvement at a broader citizen and external expert level. This was also occurring via Genome Canada's regional genome centres and their GE³LS mandates and processes.

Periodic Parliamentary and Opposition Party Involvement,
Criticism, and Committee Inquiries

If Cabinet involvement in biotechnology was problematical and episodic, so too was parliamentary and opposition party interest in biotechnology via parliamentary committee inquiries and reports. We examine here selected examples that show interesting and relevant impacts, although often seen as very indirect influences on biotech policy, governance, and evolving controversies.

An earlier example of parliamentary interest and focus centred on the use and regulatory approval of recombinant bovine somatotropin (rbST). When one of the first veterinary drugs manufactured using biotechnology entered the federal bio-food regulatory process, the rbST case generated controversy in several arenas, including Parliament (MacDonald 2000). Its promoter, Monsanto, argued that rbST use would safely supplement a naturally occurring protein-growth hormone that simulates milk-production in cows. The effect of the rbST would be to increase milk production in cows from 10 to 25 per cent from a corresponding 5 to 15 per cent increase in feed (ibid., 157). Although rbST was a veterinary drug product, a key part of the politics of rbST was that milk as a food product, as dairy interests and consumer groups argued, is nature's perfect food.

Over the period from 1990 to 2000, when rbST was under consideration, leading ultimately to its non-approval by Health Canada, an array of interests both supported and opposed it, the latter including the National Dairy Council of Canada. Space does not allow a detailed account of the important details of the nature of the coalitions of support and opposition, including consumer opposition, and U.S. impacts as well, but the rbST case did garner review by two committees: one in the House of Commons and one in the Senate.

In 1994, the House of Commons Standing Committee on Agriculture and Agri-Food examined the potential impact of rbST on the dairy industry. Concerns here were quickly linked to an rbST task force study that called for further scientific and socio-economic review. As a result, a one-year 1994–5 moratorium on rbST was announced by the Chrétien government. When controversy continued with subsequent regulatory review, the Senate Committee on Agriculture and Forestry held meetings on rbST in 1998 and questioned the legitimacy of aspects of the formal regulatory review process. Shortly thereafter in 1999, Health Canada refused to license rbST (MacDonald 2000).

It is rare for Parliament to review or have such impacts on particular bio-products being assessed for licensing, in part because of constraints due to commercial secrecy but also simply because it lacks expertise. Parliament's more normal level of scrutiny tends to be broader. For example, in May 1998 the House of Commons Standing Committee on Agriculture and Agri-Food reviewed the industry. Its members also travelled to Saskatoon where the industry was centred, and the committee issued, which, perhaps not surprisingly, was extremely supportive of the industry. Among its six recommendations were the need for greater federal support in marketing new bio-food and related products, and greater assistance for long-term R&D in bio-research (House of Commons Standing Committee on Agriculture and Agri-Food 1998, 11). It also recommended somewhat more critically that Parliament should review Canadian policy on bio-food product labelling.

Progress on the labelling front was non-existent, and a decade later, in May 2008, a private member's bill, Bill C-517, which would have conferred this right to know, was defeated 156 to 101 in the House of Commons on a free vote (the norm in private member's bills). Commenting on the defeat of this legislative proposal, Greenpeace noted, 'This vote ignores public opinion polls which have consistently shown between 80 and 95 percent of Canadians want mandatory labelling' (Greenpeace 2008).

The 1998 committee report also recommended an accommodating intellectual property framework that would foster the development of new products but 'without giving up the farmer's privilege' (House of Commons Standing Committee on Agriculture and Agri-Food 1998, 11). The latter qualifier referred to growing concerns about bio-food companies seeking greater patent-based controls on farmers that prevented them from reseeding crops except through purchase from patent holders.

Political orientations toward biotechnology are not uniform within the parliamentary arena. While the agriculture and agri-food committee remains an arena mainly supportive for bio-food and biotechnology overall, a committee such as the House of Commons Standing Committee on Environment and Sustainable Development (1996) was much more critical of both the design and focus of the Canadian biotechnology regime. This was certainly true in its report in 1996 on *Biotechnology Regulation in Canada: A Matter of Confidence*. As later chapters on bio-food and bio-health show, the concerns here were about a wider range of bio-products but also about the fact that in the initial design of the federal biotechnology regulatory regime, Environment Canada as a regulator was in the outer ring of authorities. Its jurisdiction took effect mainly if the primary regulators, the CFIA and Health Canada, did not operate.

Chapter 6 also shows the more active and long-term role of health committees in both the House of Commons and the Senate in the long, slow, and difficult development of issues, laws, and agencies regarding assisted human reproduction. These involve roles where the committees were used by ministers to test out the nature and strength of MP and voter views about reproduction technologies and eventually about related bio-life issues linked to the mapping of the human genome.

A more recent example of parliamentary debate and actions that were critical of biotechnology occurred in 2010 when an NDP private member's bill passed second reading in the House of Commons but was later defeated on final reading in March 2011. Bill C-474 was a bill to protect farmers by calling for an analysis of potential harm to export markets prior to approving new genetically engineered seeds (New Democratic Party 2010a). A further NDP private member's bill was also introduced to prohibit discrimination via changes to the Canadian Human Rights Act on the grounds of a person's genetic characteristics (New Democratic Party 2010b).

Thus, the Canadian Parliament has been a policy-influencing body within the broader context of growing executive dominance by prime ministers and Cabinets. No mere or automatic rubber stamp of government policy on biotechnologies, members of Parliament and senators have regularly raised questions, presented concerns, conveyed the views and preferences of their constituents and other Canadians, and offered alternative courses of action.

Biotech Policy and the Search for Technology Assessment Areas

A final complementary way of conveying the nature of biotech policy institutions and decision process is through a brief discussion of the larger challenges of dealing with *transformative* or *enabling technologies*, in particular the search for more formal arenas and processes for technology assessment. *Transformative technologies* refer to periods of time and development where economies and societies are transformed by new technological changes. Past examples of such developments comprise the introduction of steam power, electricity, the internal combustion engine, or nuclear power (Phillips 2007). Such technologies spawned a debate in the 1970s and since about how and whether governments can practise rational forms of technology assessment and engage in technology foresight.

The issue of transformative technologies is of fundamental importance in the current era for the simple reason that several new technologies have emerged and, equally important, are interacting, producing myriad applications as well as generating intended and unanticipated consequences for economic affairs, social interactions, and notions of human identity. At the same time, these applications and consequences become the subject of political debates and the objects of public policy. These include technologies such as information and computer technology, biotechnology, and nanotechnologies, all of which have links to and potential for generating other sectoral technologies, products, and processes in different industries and social realms (Fukuyama and Wagner 2001).

Long before biotechnology emerged as a transformative technology, the Canadian government and other states debated how to deal with technology assessment as a necessary part of democratic policy- and decision-making. The need was for institutions and processes that would enable a society and polity to assess new technologies, their benefits, and their potential adverse impacts before they emerge in the form of new products and production processes (Nordman 2004). Such issues were raised regarding the peaceful uses of nuclear energy in the 1950s and 1960s, and, in the 1970s, regarding technologies such as supersonic aircraft.

These and other issues prompted the United States to establish the Congressional Office of Technology Assessment (OTA) in 1972. The OTA published numerous studies on technology assessment but was abolished in 1995 following the Republican victory in the mid-term elections (Mooney 2005; Nye 2006). Its demise was due to sharp partisan politics and policies by a conservative-dominated Congress, which some saw as 'anti-science' as well as being opposed to government intervention of the kind that some saw in technology assessment. Technology-assessment institutions and issues continue to be examined through international network bodies, such as the International Association for Technology Assessment and Forecast Institutions and also through diverse literatures and journals.

Canada had nothing equivalent to this kind of designated technology assessment office or mechanism (Doern 1981). During this period, Canada did have broad-based science policy advisory bodies such as the Science Council of Canada and the later National Advisory Board on Science and Technology (NABST), but these bodies only episodically and very generally dealt with particular new technologies. The Canadian Biotechnology Advisory Committee was itself probably the closest example Canada has had of a technology-assessment entity – but it is gone, it too the victim of a Conservative administration – and it focused on only the one area of technology: biotechnology and related genomics research issues. A new overall federal advisory body, the Science, Technology, and Innovation Council, has recently been established by the Harper Conservatives, but it is too early to determine exactly what it will focus on in this large realm of decision-making issues (Industry Canada 2007). In recent years, the federal government also established the Council of Canadian Academies, with the view that it might help fill the kind of arm's-length analytical gap in Canada that was performed partly in the U.S. policy system by the American Association for the Advancement of Science and other bodies.

Another manifestation of the need for some form of broader technology-assessment arenas came in discussions in 2007 and 2008 regarding the need for what was labelled simply as a possible federal decision-making framework (DMF). Broadly speaking, the suggested federal DMF would be a framework or approach for dealing more systematically with the 'non-science' social and economic values, issues, and impacts of transformative technologies such as biotechnology (from whence the DMF proposal initially emerges) but also other technologies such as nanotechnology.

The DMF project discussion was seen by Health Canada and some of the federal science-based departments involved in biotechnology as an effort to deal better with the 'societal and economic issues' that accompany the introduction of new technology and thus go beyond the science-based safety questions regarding product approvals based on such technologies. Such a broader DMF was seen as a needed, but thus far missing, complement to the existing regulatory approval process for marketing regulated products based on assessing the risks and benefits of a product.

The rationale for a Canadian DMF linked to concerns about the overall integrity and international standing of the federal regulatory processes and decisions. 'To maintain its decision-making credibility both domestically and internationally, the Federal Government must ensure that the federal decision-making process is consistent and transparent to ensure both confidence in the federal regulators and in the fairness of the process and its accountability so as to avoid challenges to regulatory decisions, and that it is clearly the role of the federal government to make these decisions' (Health Canada 2007b, 3).

The policy objective of a possible DMF initiative was to develop a broader decision-making approach to 'facilitate clarity, consistency and the enabling of choices when making decisions regarding the societal and economic implications of research activities and classes of products of biotechnology. The absence of such a decision making mechanism ... has exposed federal departments to potential criticism from either the public or industry' (Health Canada 2007b, 3).

In addition, it was stressed that the DMF had to be 'internationally consistent. To provide a tool for the federal departments to use as a federal reference, the framework will build on the guidelines laid out in the Stewardship Framework and must respect laws and commitments, add value to the departments, and be clear, simple and more precise and operational' (Health Canada 2007b, 4).

A DMF, therefore, was seen in quite a multifaceted way as a decision-making approach that would be capable when necessary of signalling to business interests but also to other social stakeholders that some classes of products being developed through transformative technologies may be acceptable and others may not.

Undoubtedly, this is a difficult and complex process. Moreover, any such DMF or related technology-assessment process must be linked to existing Cabinet, parliamentary, and other policy processes already in place. Not surprisingly, no such process has emerged. But the elusive search for such processes is necessarily a part of the bio-policy and bio-governance story being told in this book.

For example, within the overall health and drug policy ambit (including bio-drugs and genetic testing products) 'technology assessment' of a certain type occurs under the work of the Canadian Agency for Drugs and Technologies in Health (CADTH 2006). CADTH is not a regulatory body by itself but rather, in the context of the 2003 federal-provincial Common Drug Review policy, develops evidence-based clinical and pharmaco-economic reviews to assess a drug's cost-effectiveness compared to existing similar drugs. These reviews are used by the Canadian Expert Drug Advisory Committee (CEDAC), an independent advisory body of professionals in drug therapy and evaluation. This committee in turn makes recommendations to the provinces on what drugs to include in the formularies of the participating drug plans. While *technology* is a part of the CADTH name, it is by no means an overall technology-assessment body of the kind implied in the United States.

More recently, Canadian policy analyst Peter Phillips (2007) has written a highly useful conceptual book on governing transformative technological innovation and raised several difficult questions about who is in charge. Phillips fittingly stresses the inherent complexity, multi-stakeholder shared relations, and competing nature of modern transformative technology governance. His conception of governance 'in practice' probes the differences in governing nominal, but highly linked stages such as knowledge, invention, the gestation of an invention, and finally, production, marketing, and consumption. Although Phillips does not fully discuss technology-assessment bodies, his framework shows some of the dilemmas in finding the right locale or 'stage' for a technology-assessment institution or process in general or, expressed in our analysis, for various bio-realms.

A related and long-standing discussion in the field of public policy institutions and practice, which emerged in the late 1960s, remains analytically and practically relevant today regarding bio-policy and governance issues. At the time, policy analyst Amitai Etzioni referred to the need in 'an active society' for a 'mixed scanning' model of decision-making – a system that he contrasted with the 'rational' model (or rational stages approach), which was frequently discussed at the time, and the 'disjointed incremental' models of decision-making (Etzioni 1968). In a sense, he combined them into his notion of mixed scanning, observing that 'each of the two elements in the mixed-scanning strategy helps to neutralize the peculiar shortcoming of the other: but incrementalism overcomes the unrealistic aspects of comprehensive rationality (by limiting it to contextuating decisions), and contextuating rationalism helps to right the conservative basis of incrementalism' (283). Two basic realities are suggested by this notion of mixed scanning: that most incremental decisions specify or anticipate fundamental decisions, and that the cumulative value of the incremental decisions is greatly affected by the underlying fundamental decisions made by a government or public authority over time.

Modern states have long been aware of the quest to find this legendary power of macro longer-term decision-making to complement and better inform the rampant incrementalism that dominates everyday decision-making. This is what much of the more explicit technology-assessment literature has been about as well. In the 1970s and 1980s, federal decision-making added more explicit 'planning' and policy branches to most federal departments and agencies, but these were not focused on technology assessment, and, in any event, they were severely pruned back during the deficit-reduction reforms of the mid-1990s (Prince 2007). In the three bio-realms, we suggest that biotechnology assessment has been a decidedly hit-and-miss process, as different kinds of interest networks sought to both promote and regulate diverse and expanding bio-realms and to use conflicting science-based and other non-science values, norms, and policy criteria.

Conclusions

Providing a basic empirical portrait of the science, research, and promotion-related agencies and bodies in the federal biotech governance regime, this chapter shows the need to see *supportive governance* occurring fairly unambiguously in the discussion of the five federal research entities, keeping firmly in mind that four of them have non-bio-research mandates and so bio-research is only one part of the research fields they support and fund or carry out. Only Genome Canada has an overall biotechnology focus in bio-health and bio-life. The analysis also shows the growing and deepening networked nature of these bodies in terms of governance, research support, and conduct of bio-science.

Within and among these research entities, biotechnology has garnered a growing research presence, eventually across all the bio-food, bio-health, and bio-life realms. Most of these demands and applications for research were not just a government creation but came from increasingly networked sets of researchers, corporations, universities, hospitals and clinics, and international competitive pressures to be a player in the bio-economy and bio-society.

We classify biotech policy and advisory bodies as support/promotion-related in that they are, or have been, part of the supportive governance structure for biotechnology taken as a whole. At the same time, they deal with, understand, and partially agree with values and interests that were critical of and/or opposed to biotechnology.

In each advisory arena, nuances in mandate and operating style differ. Industry Canada's support instincts had to be couched in the larger policy scope of the biotechnology strategies, as well as the other mandate realms its minister had within the departmental mandate ranging from consumer issues to other marketplace framework areas such as intellectual property, especially regarding patents.

CBSEC was located in Industry Canada but had to struggle continuously to engage ministers, deputy ministers, and ADMs in the intricate interdepartmental embrace of agencies with mixed bio-regulatory, support, and of course broader non-biotechnology mandates as well.

The external role and mandate of CBAC until its demise in 2007 was both crucial and valuable, but it also garnered strong criticism from many NGOs that did not really believe that it was independent or independent enough. Parliamentary committees often brought quite critical views and alternative policy ideas to bear on each of the evolving three bio-realms and thus had important impacts not borne out by the often stereotypical view that Parliament has a marginalized role in executive and business arenas. These committees were partly also an arena for the array of networked interests to seek out supportive locations for their pressure and concerns in the standing committees on health, environment and sustainable development, agriculture and agribusiness, but also in the realms of private member's bills.

Given political realities, biotechnology rarely appears at the top of the national policy and political agendas in Canada, as expressed in Throne Speeches and Budget Speeches. In reviewing these strategic documents across thirty years and several prime ministers, both Conservative and Liberal, we find biotechnology functioned at middle levels, often subsumed in discussions on innovation, health research, and broader ways of expressing national priorities. We located this relatively low-priority status in the context of challenges in dealing with transformative or enabling technologies and technology assessment. The great difficulty in developing and maintaining arenas and concepts of technology assessment includes Health Canada's effort to devise for biotechnology (and other technologies) what it calls a 'decision-making framework' conceived as an arena for dealing with 'non-science' social and ethical matters.

In the next three chapters, our analytical focus shifts from supportive governance to a style more regulatory in nature. The supportive research and promotion-related institutions discussed here wend their way into each of the bio-food, bio-health, and bio-life realms. So also do further elements of intellectual property, science-based, evidence-based, and precautionary governance emerge, along with the intricacies of technology.

4 The Bio-Food Realm: Business-Dominated Pluralistic Power

Introduction

The bio-food governance realm, the first to congeal by the early 1990s as a recognizable part of the Canadian state's biotechnology system, has evolved in the two decades since within a dynamic context of competing scientific, governmental, and political values and interests. Along with biotechnology food products that people consume, this governance realm involves broader links in the food chain to agricultural, plant, seed, and feed products. in addition to processes related to animals, animal health, and veterinary biologics. And so it is not uncommon for bio-food to be defined more broadly as agricultural biotechnology (Canadian Food Inspection Agency 2009b). Indeed, one of the pivotal reasons for changing the name of the Department of Agriculture to Agriculture and Agri-Food Canada was to emphasize the growing importance of the secondary food-manufacturing sector, especially bio-food, along with traditional primary agriculture and farm interests (Doern 2004; Skogstad 2008).

While food products primarily define this bio-realm, health and safety aspects of bio-food products are also innermost to the governance regime. At the regulatory centre of the governance structure in the bio-food realm are food and health regulatory bodies, specifically the Canadian Food Inspection Agency (CFIA) and Health Canada, through several of its agencies (Health Canada 2009d).

In this examination of the bio-food realm, our focus is on the regulatory governance aspects and on the science-based and early precautionary features underpinning it. We show how these are largely the product of business-dominated pluralistic power. Nominally, this includes an interacting array of business, consumer, science, and environmental interests and related agencies of the state. In actuality, at its core the bio-food realm is mainly a business versus consumer struggle, with bio-food business the dominant player. In part this is because of the approaches taken by Canadian producers and developers of bio-food products, in concert with U.S.-centred bio-food business interests, in the design of the international bio-food regulatory-governance system.

We look particularly at the origins and evolution of the *regulation-making* features of the bio-food realm and the nature and dynamics of two of its *product-assessment* and *approval processes* and concepts: bio-foods cast as novel foods, and plants with novel traits (PNTs). Many of the bio-food products approved to date had to secure patents through the intellectual property system beforehand. The bio-food politics have thus been influenced by the Canadian and global agribusiness forces favouring longer patent periods as property protection. These stood in sharp contrast to other NGO pressures that stressed and called for a patent system characterized by more public knowledge dissemination.

The first section focuses on overall regulation-making regarding bio-foods and how and why the regulatory system is designed the way it is, why business influence is the key feature of the bio-realm's institutionalized pluralistic power structure, and how and why it congealed. The second section concentrates on pre-market bio-food product assessment and approvals, with initial attention on bio-foods defined and regulated mainly by Health Canada under the concepts and rules regarding novel foods. The product-assessment process on plants with novel traits is then examined in the third section, where attention shifts to the CFIA. Other aspects of related bio-food and agricultural product approvals, such as animal feeds and veterinary biologics, are referred to periodically.

The fourth section looks briefly at some trends in post-market regulatory monitoring processes for bio-food products – the area historically least focused on in the bio-food regulatory realm overall but is increasing in importance because of bio-health reforms. Regarding bio-foods, this includes aspects of regulatory monitoring once a product has been allowed on the market and also crucial issues of consumer food label regulation or the lack of it. Conclusions follow about the nature and evolution of the bio-food realm, its institutionalized pluralistic power structure, and how it relates to different forms and criteria of democracy.

Bio-Food Regulation-Making: Initial Design and Congealment under Institutionalized Pluralistic Power

The bio-food regulation-making aspect of governance refers to the politics and processes of establishing the rule-making system, including its statutory mandates and underlying delegated regulations, guidelines, standards, and codes. These processes are periodic, occurring every five to ten years as new business and NGO pressures, technical developments, and international changes enter the attention span of the key agencies and their ministers, leading to new or amended rules.

Table 4.1 shows the statutory and regulatory features established by the early 1990s and announced in 1998 as the formal food biotechnology-governance structure for the 1998 biotech strategy. Its basic features are essentially the same today regarding bio-foods. Some bio-health product elements are captured in it as well, although these are left for analysis in chapter 5, when new Health Canada directorates are added to deal with bio-health products and when other federal regulators become crucial players as well, including patent agencies. It is food and related food-chain agricultural products that dominate this bio-realm. Eight different laws anchor the system, along with regulations emanating from this core statutory base (Canada 2009b; Doern 2000a; Doern and Sheehy 1999; Industry Canada 1998; Prince 2000).

As a guidepost, Table 4.1 conveys the several pathways and erstwhile stages of product assessment, depending on the nature and intended use of the product. The pathways are not just the result of laws and regulations as such, but also grow out of different inherent physical and technical realities about the nature of foods versus animal feeds, versus seeds, versus veterinary biologic products, and the like. There are also different institutional cultures in the two main regulatory bodies we are focusing on in this chapter: the CFIA and Health Canada (Doern and Reed 2000; Murphy 2006; Prince 2000).

Democratic consultations with key stakeholders and the public, though extensive and complex, have been criticized heavily by NGOs and produced strong views about how the bio-food governance realm should be established and configured. Key aspects of such consultations are governed by federal government-wide policy on regulation, including the requirement for a Regulatory Impact Assessment System (RIAS). Introduced in 1988, the RIAS requires each federal department to submit an impact statement for each new regulation and amendment

Table 4.1: Bio-food regulatory agencies, mandates, and regulations at a glance

Federal agencies and departments

- Health Canada
- Canadian Food Inspection Agency
- Environment Canada
- Fisheries and Oceans Canada

Mandate laws, policies, and regulations

- Food and Drugs Act
- Fertilizers Act
- Feeds Act
- Seeds Act
- Health of Animals Act
- Pest Control Act
- Canadian Environmental Protection Act
- Fisheries Act
- Policy Concept of Substantial Equivalence
- Food and Drugs Regulations
- Novel Foods Regulations
- Medical Devices Regulations
- Cosmetics Regulations
- Fertilizer Regulations
- Feeds Regulations
- Seeds Regulations
- Health of Animals Regulations
- Pest Control Regulations
- New Substances Notification Regulations
- Fisheries Regulations

Products regulated (examples)

- Foods, including corn, soybeans, canola, potatoes, tomatoes
- Drugs, cosmetics, medical devices (see chapter 5)
- Fertilizer supplements, including novel microbial supplements
- Feeds, including novel feeds
- Plants, including plants with novel traits, including forest trees
- Veterinary biologics
- Pest control products
- Aquatic organisms
- Products for use not covered by other federal legislation

Source: Adapted from Industry Canada (1998), *Renewal of the Canadian Biotechnology Strategy: Related Resource Documents* (Ottawa: Industry Canada), 13; and Canada (2009), Bio Basics, http://www.biotech.gc.ca.

(Mihlar 1999). The RIAS focuses on a requirement for a basic cost-benefit analysis along with other important features of pertinent information and public transparency. The policy requires that departments and agencies such as the CFIA and Health Canada publish proposed regulations and amendments in the *Canada Gazette*.

Specified forms of information are required, and opportunities for commentary by stakeholders are built into this multi-step process. Thus, a pluralist interest group approach to democracy seems inherent in the process; however, in the case of bio-food governance, the result in terms of political power is an institutionalized authority structure where business and consumer interests are the main influence bases, with the former being predominant in the design and congealment of the bio-food realm overall. The RIAS aspect of the process as well requires that the costs and benefits of the regulations as such are also estimated and revealed. Cost analysis relates only to the financial costs to business and the government of the regulatory system being proposed. Since this kind of analysis does not assess the products or the social costs, it has generated other criticisms as to how consistently it is actually carried out (External Advisory Committee on Smart Regulation 2004; Mihlar 1999).

In addition, the CFIA and Health Canada undertake their own consultations that take into account their own ways of relating to their stakeholder communities and the public. In this brief section, we note two relatively broad consultation processes that shaped and resulted in two key features of bio-food regulation: Health Canada's consultations in the 1990s that produced the current regulations on novel foods, and the CFIA's consultations, which centred on the CFIA legislation itself (when the CFIA in 1997 became a new separate operational agency) yet also extended to some regulatory change flowing from the main statute (Canadian Food Inspection Agency 2000b).

Health Canada's consultations on novel foods began in 1992 with an information letter sent to all stakeholders. This led to a formal consultation draft 'Guidelines for the Safety Assessment of Novel-Foods' in 1993 and a workshop on the regulation of agricultural products of biotechnology also held that year. Publication of 'Guidelines for the Safety Assessment of Novel-Foods' followed in 1994 as well as the formal *Canada Gazette*-based process (Canadian Food Inspection Agency 2000b).

A pre-publication phase launched in 1995 resulted in thirty-five responses from stakeholder groups, which ranged across the full

spectrum including consumer advocacy groups, provincial governments, food and biotechnology companies, academics, industry
associations, and standards organizations. Later, in 1998, a second
pre-publication *Gazette*-based phase occurred, eliciting nine responses.
Overall there were numerous formal and informal meetings, written
and verbal communications, and proposals and guidelines posted on
Health Canada's website. The RIAS element raised issues regarding
not only the details and adequacy of the consultation process but also
resulted in clearer and narrower definitions of *novel food* and of what
constituted *major change*. Comments were also received about the extent
and nature of regulatory benefits, costs, and burdens (Doern 2000c).

The points above reflect a lengthy and complex process across virtually a whole decade that raises obvious questions. For instance, the
fact that there were thirty-five groups involved initially that dropped
to nine can be interpreted in a least two ways. The twenty-six groups
that ceased to respond formally might have been satisfied with their
input. Or, despite the fact that Health Canada does fund some groups'
involvement, they may simply not have been sufficiently funded to be
able to play the 'long game' of consultation that requires continued
investment of time, expertise, and other resources to participate. These
consultations were about novel foods, which, as we see further below,
is actually a larger regulatory realm than bio-food products by itself.
Moreover, Canadian and comparative assessments of bio-food politics show that the deal to have bio-foods examined as novel foods had
already been made and its forging was undoubtedly due to global business power, led mainly by the lobbying of U.S. biotech firms, and fully
supported by Canada's emerging bio-food firms as well (Doern 2000c).

The CFIA's consultation activity has also been varied and extensive.
The storyline for the CFIA necessarily extends to its pre-1997 situation when some of its component parts resided in what was then the
Department of Agriculture (now Agriculture and Agri-Food Canada).
Over the years, the seeds, feeds, and animal health elements had fashioned their own ways of dealing regularly with their specific stakeholders and with the public (Canadian Food Inspection Agency and Health
Canada 2000; Prince 2000). Consultations regarding plants with novel
traits and novel feeds can be traced to 1988 and took the form initially
of work through advisory committees and a broader consultation in
1993. They then extended to the eventual *Canada Gazette* phases, including a pre-consultation phase, which involved over two thousand stakeholders. Furthermore, in its public consultations the CFIA appeared

frequently before parliamentary standing committees between 1995 and 2000.

Health Canada and the CFIA are aware that they are judged as regulators not only by *what* they regulate but also on *how* they regulate biotechnology and also across their very wide regulatory mandates. Additionally, other crucial self-regulatory realms and institutions are involved in the bio-food governance realm, including public and private laboratories, standard-setting bodies and associations, and international rules, some of which emerge in our discussion below of plants with novel traits and bio-food product labelling.

In the initial forging of the regulation-making regime for bio-food, the roles of environmental NGOs and of Environment Canada were very active in democratic terms but marginal in terms of power, including within a system of institutionalized pluralist power. Environmental NGOs were certainly involved in different aspects of the above consultations, and Environment Canada was lobbying from within the federal government (Abergel and Barrett 2002; Barrett and Abergel 2000). Both these interest groups and departmental officials were pressing for Environment Canada to be the lead biotechnology regulator, rather than the residual one it turned out to be. Environmental interests did not succeed, in part because of the tenacity and power of the bio-food and agricultural interests and their allies in Agriculture and Agri-Food Canada, and also because Environment Canada's first two decades as a department were focused on other evolving environmental priorities (Doern and Conway 1994).

A greater environmental and Environment Canada presence emerged in 1999 when the Environmental Protection Act was amended to widen the coverage of requirements for 'new substance' notifications to include some biotechnology products. Also, as we see below, the CFIA itself was itself given the mandate to be the 'environmental' regulator for plants with novel traits.

In the actual forging of the Canadian bio-food governance realm, environmental interests were marginal, although they did make their presence felt in general debates and regarding particular bio-food product proposals. As well, we will consider the empirical issue of whether pressure from environmental interests in Canada and internationally *prevented* the approval of some such products – an issue that requires an appreciation of how to treat non-events in the form of bio-food products withdrawn from assessment processes or never even submitted because of cumulative pressure and consumer opposition or feared opposition.

Pre-Market Bio-Food Product Assessment and Approvals under Novel Food Concepts and Regulations

Bio-food products are regulated by Health Canada under the provisions of the novel foods regulations, the concept of novel foods, and the policy of *substantial equivalence*. Health Canada defines novel foods as 'foods resulting from a process not previously used as a food; products that have never been used as a food; or foods that have been modified by genetic manipulation' (Health Canada 2009d). Over seventy novel and genetically modified (GM) foods have been approved for sale in Canada, up from the total of forty-three approved by 2000. In both those figures, by far the largest component is novel bio-food products. This includes some 'stacked trait' crops, but second- and third-generation GM varieties such as GM animals await assessment and final decisions (Doern and Phillips 2012).

An initial sense of the volume of products over time is important in any analysis of regulatory governance, because volumes range widely with regard to bio-food overall and also even more starkly high volumes in bio-health products. Volumes, combined with the scientific complexity of the products being assessed, help determine the essential rhythm of the regulatory agency involved. Also, they are crucial factors in understanding the nature of the core front-line relationships and workloads of the bio-food applicant/proponent, on the one hand, and the science assessors within Health Canada, on the other. Since this front-line set of relationships and interactions is crucial to bio-food product approvals, we need to understand it with multiple dimensions in mind.

Accordingly, we look briefly first at major aspects of novel food assessment overall and then at the safety assessment elements of bio-foods. On the overall novel food assessment process, Health Canada notes the following aspects:

- *The host organism.* In order to assess the substantial equivalence of any novel plant, it is imperative that the evaluator has detailed information about the natural history of the non-modified host plant. The biology of each of the major crop species in Canada has been reviewed and published by the Plant Biotechnology Office of the Canadian Food Inspection Office of the CFIA.
- *The donor organism.* Information about the natural history of the donor organism is required, particularly if the donor or members of

its genus naturally exhibit characteristics of pathogenicity or toxin production, or have other traits that affect human health.

- *The modification process.* A detailed exposition of the molecular characteristics of the novel plant is required in order to demonstrate that the developer has critically analysed the plant and its products, including the novel genes and novel proteins. The method by which the novel traits are introduced into the host plant determines, in part, the information requirements for the assessment of the molecular biology of the plant.

- *DNA analysis of the plasmid backbone.* Agrobacterium-mediated transformation of plants is the most commonly used method for introducing novel genes into the plant genome. It results in the insertion of single, or often tandem copies of the DNA cassette as delineated by the left and right border repeats of the T-DNA. Sequences from outside of the left and right borders of the T-DNA may also be integrated along with the T-DNA, therefore the applicant must determine if any such plasmid sequences are present in the host plant genome.

- *The genetic stability of the modified organism.* The inheritance and stability of each introduced trait that is functional in the transformed plant must be determined. For each novel trait, the pattern and stability of inheritance must be demonstrated as well as the level of expression of the trait. If the new trait cannot be measured directly by an essay (for example, ELISA), then the inheritance of the new trait will have to be determined by examining the DNA insert directly, and the expression of the RNA.

- *Expressed material/effect.* The transcription and/or translation products of a novel gene, or genetic element, that has been introduced into the plant genome, or these same products arising from a modified endogenous gene, or genetic element, must be characterized. Where the result of the modification is the expression of a novel protein, or polypeptide, this material must be characterized with respect to identity, functionality, and, where appropriate, similarity to products from traditional sources. In cases where the modification is the expression of a novel non-translatable RNA transcript, the sensitivity and specificity of the desired action should be established. Examples include the production of anti-sense mRNA or other RNA species resulting in the reduced production of an endogenous protein. The altered regulation or expression of non-target host genes should be investigated in the course of assessing the safety and nutritional acceptability of the food products from the modified plant (Health Canada 2000a).

With respect to the safety assessment of bio-foods, Health Canada requires the applicant to provide data and analysis for its science evaluators regarding:

- molecular biological data describing the genetic change;
- nutritional information about the novel food compared to a non-modified food of the same type;
- potential for production of new toxins in the food;
- potential for any unintended or secondary effects;
- key nutrients and toxicants;
- major constituents, such as fats, proteins, carbohydrates, and minor constituents, such as minerals and vitamins; and
- microbiological and chemical safety of the food (Health Canada 2009d).

While these features of bio-food product assessment and approval are a crucial starting point, they do not, in themselves, convey a sense of the stages involved or of the dynamics of science-based and precautionary governance at the product level.[1]

At *the pre-application stage*, for example, are any number of discussions and exchanges between potential applicants and Health Canada staff and assessors about the data required. These early discussions are vital to the firm in that it develops its product knowing that regulatory approval (that is, no objections) is crucial to product acceptance in the marketplace. At the same time, the firm has its own internal reasons for wanting good science *within* the firm to underpin its products. These reasons centre on commercial and competitive pride in having a good efficacious product, and also on a healthy fear of future legal liabilities if the product is unsafe or ineffective.

Once an application is *received* at Health Canada, often consisting of several volumes of data, it is assessed by a multidisciplinary team of scientists, variously specialized in the core safety and data realms of science – molecular, compositional, toxicological, nutritional, and allergenic.

The *front-line core of science officers* with master's or doctoral degrees who are engaged in assessments is remarkably small and absolutely crucial to ensuring bio-food safety. Approximately ten experts in Health

Canada and ten in the CFIA (for its aspects of biotechnology in plants, feeds, seeds, and animals) anchor the safety assessment as each application comes in. These front-line science officers draw on an extensive published peer-reviewed scientific and technical literature as well as expert reports and international guidance documents, in addition to frequent, often daily contacts with fellow front-line science assessors in the regulatory agencies of other countries, particularly the United States.

The *assessment process* is initially quite reductionist. Individual assessors first look at aspects dealing with their particular expertise and then the process is augmented and aggregated by group discussion as the small assessment team analyses and debates the adequacy of the applicant's data and their implications.

Where the review team believes there are *clarifications needed* or deficiencies in the data and studies present, these are requested and required from the proponent firm. If such further information is not supplied, the application process stops and indeed the regulatory clock may be turned back to zero. The review process for each product application can take from twelve to eighteen months in Health Canada's novel foods assessment process, but there are other specified time periods and 'time clocks' during which certain aspects of regulation and information have to occur. Health Canada's time performance record has improved in response to industry pressure (External Advisory Committee on Smart Regulation 2004) – an issue we return to as an overall bio-governance regulatory issue in later chapters.

If they occur, *disputes* of either a technical or a procedural nature must be resolved or referred to an adjudicative committee. These differences must essentially be talked about and resolved partly on a professional-collegial basis among the teams within Health Canada and the applicant firm. In part then, this form of dispute resolution functions on trust and mutual professional respect for complementary and separate realms of expertise. All regulatory systems must function somewhat on the basis of trust, but none can gain public legitimacy on this basis alone.

More formally, a Food Ruling Committee meets each month to deal with many different kinds of food product decisions. Individual applications may have been discussed up the line as needed with directors general, who are also typically scientists and have usually been front-line science assessors themselves. The Office of Food Biotechnology at Health Canada receives the comments from the various assessors and prepares a recommendation, which is taken to the Food Ruling Committee.

The formal pre-market *process ends* when Health Canada advises the applicant that it has 'no objections' to the product. There are few instances of formal or official 'rejections' of a biotechnology product, but there can be de facto withdrawals from the product assessment. Such withdrawals are certainly not a formal stage in the process but could arise simply if the proponent does not submit requested data or if assessment showed that some aspect of assessment was not met. Its application assessment would simply cease, either totally or until the proponent decided that it would comply.

Health Canada's final decisions about a novel food product, indicating that the regulator has no objections, are communicated in writing to the proponent company. But the other aspect of communicating final decisions centres on communication to the public. Final decision documents on novel food products are put on Heath Canada's website, but the content of these documents does pose a challenge regarding how much detail and information needs to be in the decision document. Documents too long and too technical can generate criticism from the public. Any shortening of the document, however, can lead to the critique that they are not revealing vital information.

A further factor of considerable import is that the document must be crafted so as not to reveal commercial information that would, if published, be valuable to the proponent firm's competitors. This constraint entails issues of commercial secrecy and privilege and how the lines between public and private are themselves determined, including their links to freedom of information and privacy laws.

Alongside bio-food product assessments, another bio-food realm regulatory aspect concerns the meaning and application of the *precautionary principle*, the concept of *substantial equivalence*, the nature of science assessment and causality, and who is doing the assessing, front-line assessors in regulatory bodies versus the broader scientific community employing peer review (McHughen 2002; Millstone, Brunner, and Mayer 1999; Phillips 2002; Phillips and Wolfe 2001).

The precautionary principle tends to be enunciated and advocated first and most often by Environment Canada, one of the four departments in the federal biotechnology regulatory system; by domestic environmental NGOs; and, of course, by international environmental pressures and agreements. Also, the precautionary principle enters the bio-food realm through Health Canada and health NGO advocacy and lobbying, initially because of the Krever Commission and controversies over the regulation of blood (Saner 2002). More specifically, an expert panel study by the Royal Society of Canada (2001) explicitly called for more elements of precaution to be built into bio-food regulation.

The Royal Society's numerous recommendations included strong advocacy for the replacement of reliance on 'substantial equivalence' as a decision threshold with rigorous assessment of the bio-food's potential for causing harm to the environment or to human health. In concert with this view, the Royal Society expert panel recommended that a precautionary regulatory assumption apply, because safety assessments of GM products 'on the basis of superficial similarities' is not a 'precautionary assignment of the burden of proof' (Royal Society of Canada 2001, x). To date, these recommendations have not been adopted by the government of Canada.

The Royal Society panel report voiced strong objections to the fact that a more open peer review system is not an explicit, more public part of science-based regulation and product assessment, since peer review would, in its view, be a better public interest–centred way of determining scientific objectivity (Royal Society of Canada 2001, 213–14). It was also critical of the belief that genetic testing and inserting genes is a precise technique, particularly when it came to assessing long-term effects. In the panel's view, it is not and hence again the need for both precaution and peer-reviewed science.

However, Health Canada's novel food and biotechnology regulators tend to argue in response that their overall system for assessing products is itself a practice of precaution, even though there is no mention of the concept in its statutes (although a form of the precautionary regulatory assumption is explicit in environmental laws administered by Environment Canada, such as the Canadian Environmental Protection Act). In other words, precaution is said to apply because a product is not approved if some safety issues have not been adequately addressed and because of the careful and deliberate science-based nature of the safety assessment. This is a claim we look at again in chapter 7 after all three bio-realms have been explored in the context of the above linked issues. This means taking into full account the fact that science-based risk regulation now involves balances among science, technology assessment, related kinds of evidence, and public policy expressed through statutes, regulations, processes, and norms to detect what Sparrow (2008) calls 'the character of harms.'

A closely related question naturally arises as to whether bio-product assessments are, or should also be, influenced by non-science-based norms lumped under the label of *other socio-economic criteria and values*

(Health Canada 2007b; Nuffield Foundation 1999; Royal Society of Canada 2001). And this question, in turn, raises the issue of how and when economic and social considerations should be a part of any product assessment.

Broader stakeholder and citizen involvement deploying socio-economic criteria in product assessments by itself are rare, as our discussion of the rbST case showed in the previous chapter. But other product-specific scenarios and situations are not difficult to imagine. It may be that some of the scientist-assessors employed by the regulator have strong biases *as citizens* and individuals, rather than as civil servants, for or against biotechnology products. This is a possibility, although the notion that *teams* of scientists actually assess a product should be one powerful antidote to minimize this possibility. There is also the prospect of an institutional bias in the regulatory system that becomes pro-business or pro-speedy approval of products.

As well, the actual or expanded use of socio-economic criteria necessarily links to the issue of whether broader public consultation can be or should be built into the *product* assessment process as opposed to, or in addition to, the general biotechnology regulation-making process we discussed above. This connects with a host of important principles, which include ethical standards, precautionary criteria, technology assessment ideas, and norms regarding individual consumer choice (Middleton 1998; Wiles 2007). It also connects with somewhat contending approaches to democratic governance: democracy defined or practised as representative parliamentary democracy; federalist democracy, interest group pluralism; civil society democracy; and direct democracy.

While the bio-food product approval record seems to be moderately steady, there are also global and national issues about democratic and power structure influences on new types of bio-food products that generate new types of risk and thus new spurts of democratic concerns and opposition.

One such example centres on GM salmon. It is currently before the U.S. FDA accompanied by critics who have already labelled it 'frankenfish' (Egan 2011; McGreal 2010; Pollack 2010d). It may later be the basis for an application to Canadian regulatory authorities. Interestingly, the salmon in the U.S. application were engineered in Canada. Also of interest is that the fact that in the United States the FDA GM application for a GM animal is being examined and possibly approved under provisions regarding veterinary drugs (Leeder 2010).

Clearly the dynamics of GM product approvals is different and changing in varied political contexts, such as the regulatory climate in the United States for Roundup Ready alfalfa, a Monsanto product that resists a chemical used to kill weeds has been changing (*Economist* 2011). The product has been allowed to be used as a safe product for many years but the debate now concerns whether to set strict rules on the extent of planting allowed. In part, the concern is how GM crops can be allowed to exist beside conventional and organic ones (ibid.). Another example is in the European Union, which has historically been opposed to bio-food and has strong labelling laws. However, in 2010 the European Union approved a GM potato for cultivation, despite traditionally strong public opposition (Kanter 2010).

Product Regulation on Research on Plants with Novel Traits and Potential Unconfined Environmental Release

How research on *plants with novel traits* is regulated by the Canadian Food Inspection Agency (CFIA) is the second product realm we explore here. A PNT is defined by CFIA as 'a plant that contains a trait which is both new to the Canadian environment and has the potential to affect the specific use and safety of the plant with respect to the environment and human health. These traits can be introduced using biotechnology, mutagenesis, or conventional breeding techniques' (Canadian Food Inspection Agency 2009b).

In the past twenty years in Canada, close to 8,000 confined field trials have been conducted for plants with novel traits. Information from the CFIA reveals that from 1988 to 2000, about 5,500 confined field trials were conducted for plants with novel traits, while from 2003 to the end of 2009, 720 submissions for confined field trials were submitted involving 2,430 field trials (Canadian Food Inspection Agency 2000b, 2009c). Of the 2,430 field trials, over half were submitted between 2007 and 2009, indicating a considerable spurt of growth. But, as of 2008, only about fifty PTNs and/or novel livestock feeds derived from plants have been approved for commercial use in Canada. Not all PTNs are biotechnology-derived but most are, so these figures indicate an especially large volume of early-stage research on plants that can later potentially show up as novel food applications, feed product applications, and applications for unconfined release into the environment. Moreover, these trials are spatially dispersed across Canada and therefore require at CFIA a significant inspection front-line scientific and technical capacity.

The CFIA, in its mandate, is largely an inspection operation (Prince 2000). About 500 of its staff are based in Ottawa but the 4,100 other staff in the regulatory front lines are in the agency's field offices. Since the CFIA is inspecting the many different aspects of its mandate, its inspectors are engaged in a multitask mode of inspection. Distinctly conscious of the need for the continuous training and education of its inspectors, the CFIA runs 'inspection schools' regularly.

CFIA's responsibilities overall regarding PNTs include the approval of unconfined release of PNTs; approval and inspection of confined research field trials of PTNs; assessment of import applications for PTNs; as well as crucial involvement in both national and international development of regulatory policies related to the environmental release of PTNs. The primary trigger for assessment is the novelty of the product rather than the specific means by which it is produced. Regulators are looking for plants with novel traits, not specifically plants modified by recombinant DNA techniques. And again the concept of *substantial equivalence* is a starting point for assessment. Before allowing trials, the CFIA requires the product developer to submit the following information regarding research trials, each of which is limited to one hectare:

- a detailed map of the field trial location, including crops and planting dates;
- the presence of endangered species at the field trial location;
- the size and number of field trial locations;
- plan for the maintenance of reproductive isolation (e.g., isolated from other plants of the same or related species by specified minimum distances based on the known biology of the plant species);
- disposal of plant material (e.g., preventing the spread of PNT material into the wider environment through deep burial and complete incineration); and
- post-harvest land use restrictions (Canadian Food Inspection Agency 2009b).

The sequence of regulatory assessment proceeds through the stages indicated below (contained use, confined research field trials, unconfined environmental release and/or use as a livestock feed; and commercialization), with each stage having potentially different self-regulation, regulators, or shared regulators as follows:

- *Contained use* (laboratory, greenhouse) self-regulated by firms, laboratories, and/or through ISO certification, including rules by research-granting bodies;
- *Confined research field trials* (reproductive isolation) regulated by CFIA's field testing assessment;
- *Unconfined* environmental release (reduced or no reproductive isolation) safety assessment by the CFIA; and (if above trials are approved) then
- *Commercialization* (variety registration and feed use) regulated by the CFIA and food use by Health Canada (Canadian Food Inspection Agency 2000b, 2009c).

The initial *contained use* stage raises issues about self-regulation and how confident the public can be about its transparency, efficacy, safety, and compliance. At the same time, however, it is difficult to envisage how this aspect could be anything other than dependent on significant forms of self-regulation, given the numerous sites and locations for research and field trials.

Confined use field trial rules specify the information requirements listed above. The CFIA inspects all trials. This means that the regulatory and compliance process is crucially dependent on accurate records being kept and on a visible expert presence in the field. Cost-recovery secured from the applicant firm or product developer has been introduced by the CFIA to fund this aspect of the regulatory process.

Regarding the *confined research field trial* sites, particular controversy arises because in Canada there is neither mandatory public notice about the trials nor any information about the exact location of the trials. A thirty-day notice is given to the provincial governments involved. The absence of wider public notice and information on exact locations is largely due to two factors. First, there is a fear of damage to the site, which may invalidate tests. But second, keeping in mind extraordinarily public anti-GM food and plant protests in the United Kingdom (where sites were revealed), there was simply a desire to avoid such front page and media opportunities to the critics of the GM food industry (Nuffield Foundation 1999; Royal Society of Canada 2001).

Also there is a detailed *unconfined environmental release* safety assessment in the Canadian governance realm. At this phase, CFIA evaluators assess a number of features and criteria, again based on volumes of complex scientific data submitted by the product developer. Examples of the questions posed and decision factors considered include:

- Does the plant have the potential to become a weed of agriculture?
- Is there potential for gene flow to wild relatives whose hybrid off-spring may become more weedy or invasive?
- Does the plant have the potential to become a plant pest, or have an impact on an existing plant pest?
- Is there a potential impact of the trait on non-target organisms?
- Does the plant, or its products, have the potential to affect livestock feed or food safety? (Canadian Food Inspection Agency 2009b).

When a CFIA assessment is completed, a decision document is sent to the product developer and later published on the CFIA website. These documents explain the conclusions reached and provide details on the basis of the decision.

Following an unconfined release approval, generally there are no requirements for reproductive isolation, site monitoring, or post-harvest land-use restrictions; an exception is for Bt insecticidal protein expressing plants, where resistance-management plans are imposed (Canadian Food Inspection Agency 2000b, 2009b).

On the process overall, a final point is how the 'environmental' aspects are assessed in the broader confined and unconfined stages. The concerns here are scientific and jurisdictional. Scientifically, the concern is whether the process of assessment allows for the examination of cumulative and long-term impacts (National Research Council [U.S.] 2010; Nuffield Foundation 1999; Royal Society of Canada 2001). Field trials are not assessed through public consultation in the way in which Environment Canada's Canadian Environmental Assessment Agency conducts its environmental assessment of federal projects under its legislation. But the CFIA does assess wide environmental aspects in this research/trial phase, and environmental assessment is central to its inspections and enforcement.

The core front-line relationships between CFIA science assessors and personnel in the product developer are in many ways similar to the novel food/bio-food process. Assessments depend on diverse technical disciplines, and assessments by groups of assessors are crucial. One key difference from novel foods is that the volume of assessments, as noted above, is much greater. There are more diverse product developers that include firms and universities, and there are the spatial imperatives that arise from diverse field trial sites and eventual possible environmental release areas located across Canada.

Bio-Food Post-Market Regulatory Monitoring

Strategic aspects of this phase of bio-regulatory governance centre on concerns about bio-foods, plants with novel traits, and consumer labelling of bio-food products once they are on the market. Once a biotechnology *food* product approved as a novel food is on the market, there is no systematic post-market monitoring of the products by the regulators. In other words, there is no systematic program of 'taking another look' at a product a few years later. Nor are there publicly identified ways set out by Health Canada of simply reporting further on product performance or possible adverse effects, if any. Companies that hold licences to sell bio-food products are under a legal obligation to advise Health Canada of any adverse effects of their products of which they may become aware. Health Canada, however, does not reveal the extent to which this is being done or what actually happens or does not happen afterwards.

This situation does not mean there are no avenues through which post-market concerns and information might be brought *informally* to the attention of regulators. Citizens can always in some way simply contact the regulator. But *formal* post-market review of biotechnology novel foods has not been a strongly explicit or well-resourced and staffed aspect of the bio-food product-governance regime in Canada.

A study by the National Research Council in the United States (2010) points to the continuing need for objective arm's-length scientific assessment of the overall use and impacts of bio-food crops. The scientific and political significance of this study is that it is arguably the first comprehensive assessment of how such crops are affecting all American farmers, including those who grow conventional or organic crops. Perhaps not unexpectedly, conclusions from the study show mixed outcomes of positive results along with negative consequences.

The positive environmental effects centre on considerable improvements in water quality, and there are also economic benefits in reductions in the costs of production and the reduced use of pesticides. Nevertheless, the study also points to concerns still about the long-term effects of some bio-crops and about the need for more sustained science-based independent monitoring. Arguments about further reform of this kind have also found their way into U.S. analysis because of the need to factor in new understandings of the networked nature of genes (Van Tassel 2009).

Another post-market economic evaluation examined maize (corn) yields for fourteen years in the United States and found that 'farmers who continued to grow conventional crops actually earned more money . . . than those who cultivated GM varieties' (Connor 2010b). The non-GM farmers benefited financially because they did not have to pay the extra costs of purchasing GM seeds. But both sets of farmers benefitted from the lower level of pests in large part, according to the evaluation, because non-GM farmers became free-rider beneficiaries of the pest-reduction impacts from nearby farmers using GM maize.

Broader health product–governance changes in post-market monitoring may yield changes on the bio-food side as well. Not all post-market situations, it warrants mention here, are the same. For example, drugs have toxic effects that are better known, and thus comparisons with novel foods may not be fair regarding the extent of post-market review possible or needed.

With respect to PNTs, the post-market regulatory story is quite different because of the distinct nature of PNTs as a 'product' and because of the CFIA's role as an enforcement agency. Any authorization for a PNT and/or novel livestock feed derived from plants is subject to a CFIA review, if information comes to light about new or changed unforeseen risks, and these must be reported by the person or company involved.

In addition, the CFIA conducts a range of post-approval inspections and monitoring to verify that all registered products continue to meet quality and safety standards.

Assorted forms of follow-up review occur after research field tests have been conducted:

- Field trials are monitored by CFIA inspectors.
- Field trial records can be requested by agency inspectors at any time.
- Post-harvest is monitored by agency inspectors with time periods dependent on species.
- Seed harvest records must be kept and methodologies for excess seed disposal must be provided.
- Developers/applicants/growers are legally bound to report any unforeseen events or new information to the agency.
- Where appropriate, pest resistance–management schemes must be submitted and can be audited at any time by agency inspectors,

such as Bt potatoes and corn (Canadian Food Inspection Agency 2009; Canadian Food Inspection Agency and Health Canada 2000).

A final crucial post-market topic in the food biotechnology regulatory system is consumer *information and choice*, an issue that illustrates the political intersection of the bio-economy and bio-society. Health Canada and the CFIA have joint responsibility for food labelling, the former dealing with health and safety matters for bio-foods and the latter dealing with non-bio-novel foods. The CFIA's labelling mandate deals with ensuring that food labels are truthful and not misleading.

Canada's current bio-food labelling regime remains a voluntary labelling system – the result after two decades of lobbying and conflict between health, consumer, and environmental NGOs and bio-food producers and other agricultural interests. In 2004, the federal government announced the official adoption by the Standards Council of Canada of the Standard for Voluntary Labelling and Advertising of Foods That Are and Are Not Products of Genetic Engineering, as a National Standard of Canada (Canadian Food Inspection Agency 2009b). The standard was negotiated and kept as a voluntary measure through a five-year discussion by a multi-stakeholder committee facilitated by the Canadian General Standards Board and then reviewed by the Standards Council of Canada, the administrative body that oversees Canada's National Standards System. The involvement of these entities and processes shows the importance of co-governance rather than just regulatory governance by the state.

In particular, the standards allow manufacturers, including those of bio-foods, to voluntarily use labels that tell consumers how a food is produced. The CFIA's website material on labelling concludes that Canada's combination of legislative and standards-based approaches, along with information through other sources – such as Internet and toll-free information lines – has the potential to provide consumers with more information than is available to consumers in countries that have simply adopted mandatory labelling policies for GE foods (Canadian Food Inspection Agency 2009b).

This optimistic but qualified conclusion about the 'potential' for consumer benefit and choice reflects the fierce battle from the outset over bio-foods as 'Frankenfoods' and overt rights to consume foods that consumers know are natural and not genetically modified (Knoppers and Mathios 1998; Wiles 2007). For such interests, the 2004 voluntary standards were a meagre return on their lobbying investment. And a private member's bill in 2009 that would have required compulsory labelling fell to defeat, despite a free vote for all members in the House of Commons.

Most consumers of biotechnology products, especially in North America, are probably unaware that they are consuming biotechnology food products. And Canadian and American supermarket chains have not made such kinds of choice a central part of their competitive strategies for market share. In the United Kingdom, whose food supermarket industry is more competitive structurally, some major supermarkets initially positioned themselves in the marketplace on the basis of saying that they are selling only GM-free products. Often, however, in complex food production chains, it is difficult to know if such assurances can be independently validated or there may be problems in obtaining GM-free supplies.

Recent U.K. developments may foreshadow issues of wider importance on this front. A 2009 report by the U.K. food regulator and the environment, food, and rural affairs department stressed that U.K. supermarkets were finding it more and more difficult to find non-GM supplies (Hickman 2009). This was because major producers, including American and Brazilian suppliers, were switching to GM and so supermarkets were now paying 10 to 20 per cent more for conventional supplies. The report furthermore noted that concern about GM foods voiced by consumers had fallen steadily from a peak in 2003, when 20 per cent were worried, to 6 per cent in 2009. The CEO of Tesco, the United Kingdom's largest supermarket company that had gone GM free was prompted to add that 'it may have been a failure of us all to stand by the science' rather than giving in to the concern about GM foods (qtd in Hickman 2009). This CEO also linked GM food and higher food prices to the need to appreciate that those same GM foods could play a vital role in feeding the world's growing population. Not surprisingly, NGO and consumer interests opposed to bio-foods in the United Kingdom expressed strong disagreement with this view and with the main arguments of that report.

As a final cautionary point, we want to stress that bio-food politics and governance are often buffeted, positively or negatively, directly or indirectly, by other forces and controversies – issues and factors that may be underway in national and international arenas of food safety governance overall. For instance, in 2008 and 2009, deaths due to an outbreak of listeriosis prompted a review of the CFIA and Health Canada (Canada 2009b). This review led to proposals that included changing the CFIA governance structure to a more representative board structure as existed in the United Kingdom and the European Union. Other food-borne disease outbreaks have generated similar serious political reviews in the United States, United Kingdom, and Australia. While bio-food was not central to these recent reviews/controversies, the post-review processes could lead to structural reforms, as well as reforms on the post-market side of food safety, which will affect bio-food governance (Doern 2010).

Conclusions

The bio-food realm was the earliest one to consolidate as a recognizable part of the biotechnology governance system in Canada and also in the United States. Regulatory, science-based, and precaution-based aspects of the governance system have been our focus in this chapter, especially on the roles of Health Canada and the CFIA. We characterize the design and nature of Canada's bio-food realm as occurring under a system of business-dominated pluralist power. Pluralist consultations are often used, but the system is institutionalized in a way that favours and reflects business dominance, particularly in the central battle between business and consumer interests. Business interests play the commanding role, as demonstrated by the systematic failure to achieve compulsory labelling of bio-food products, as had taken place in Europe.

We have looked at the origins and evolution of both the regulation-making features of the bio-food realm and the nature and dynamics of two of its product assessment and approval processes and concepts: bio-foods assessed as novel foods and plants with novel traits.

The regulation-making aspects of the bio-food realm reveal considerable complexity regarding the number of statutes and regulations, and hence their potential linked product cycles that underpin this realm of bio-governance. However, it also tells a simpler compelling story of institutionalized pluralist power in that the system was designed to focus on the primacy of the core health and food regulators, with the intended effect of excluding and pushing to the margins environmental regulators such as Environment Canada.

Though considerable consultation and debate have accompanied the regulation-making and design aspects, there is little doubt that bio-food business interests shaped most of the bio-food system to suit their needs. These needs were twofold: to have bio-foods treated as a particular kind of novel food or novel plant linked to the policy concept of substantial equivalence, and to gain legitimacy for product efficacy through effective front-line science assessment and fairly speedy and efficient regulation.

For their part, Health Canada and the CFIA both saw the initial regulatory regime design bargain in terms of public interest health and safety, with a particular emphasis on pre-market product assessment and safety. Health Canada and the CFIA also had a particular appreciation of the need to align the Canadian biotechnology regulatory approach with international governance ideas, including larger novel food and novel plant concepts, together with such norms as substantial equivalence. As a result, they have more often than not resisted those interests – NGO and scientific, such as the Royal Society of Canada – that sought to replace the substantial equivalence concept, extend precautionary features for both environmental and health effects, and use peer-reviewed science as the standard approach rather than science-based regulation. We return to these crucial gaps and the issues they raise in the final chapter.

In analysing pre-market product assessment, we emphasized the need to appreciate the different technical features of the products, and the assessment dynamics, stages, volumes, and front-line science evaluator–applicant relationships. The multi-disciplinary and group nature of the science-based regulatory governance assessments also clearly emerge. The CFIA's mandate and role regarding PNTs showed the complex assessment stages for high-volume research and field trials leading to possible unconfined environmental release. While the CFIA's enforcement responsibility appears formally large here, regulation is also performed by self-regulation entities at different research sites and product development stages.

Our analysis suggests a need to recognize the diverse senses of what *bio-products* are. Ultimately, they are products that reach the marketplace of course, but large-volume processes and mandates also relate to distinct 'research' activities/trials as de facto 'products' for the regulators concerned. It is in these initial and intermediate stages of the governance realm's regulatory cycle that the actual nature and likely effects of a bio-product are debated and authoritatively determined.

With respect to post-market regulatory monitoring after products are approved, again there are differences in the bio-food product cycles and those of PNTs. Post-market monitoring is more pronounced and present in the latter than in the former. Chapter 5 takes up the stronger recent pressures for greater emphasis on the post-market phase, particularly as new bio-health products come on stream. The regulation of bio-food is likely to be affected by these regulatory governance changes, emanating mainly from Health Canada.

The issue of consumer labelling and choice is a particularly contentious one for the bio-food realm. Bio-food business pressure has limited bio-food labelling to a voluntary feature in Canada. NGO and parliamentary committee pressure to make it a mandatory feature of post-market bio-food regulation has not succeeded, as it has in Europe. The bio-food realm still anchors the overall Canadian bio-governance regime, located within the larger milieu of research and support/promotion-related agencies and the larger changing policy architecture. From the mid-1990s on, however, biotechnology as an enabling technology began changing quickly, as the next two bio-realm governance chapters on bio-health products and on bio-life amply demonstrate.

5 The Bio-Health Realm: Networked Power and Governance

Introduction

Issues of population health feature in all realms of biotech policy and governance. The previous chapter has shown there are crucial health aspects to food regulation and bio-food product approvals, and in the next chapter we will illustrate how questions of health and medical science are integral to the bio-life realm. Here we are concerned with the bio-health realm itself and suggest that with intensifying new bio-health research and bio-health products, especially since 2000, the bio-health governance realm has a clearer presence and a growing part of the Canadian biotech governance regime.

Our basic argument in this chapter is that the bio-health realm is very much in flux because of explicitly wider and more complex views of what health product governance taken as a whole should be like, not just bio-health products. Health Canada's *Blueprint for Renewal* initiatives centred on a life-cycle regulatory and post-market pharmaco-vigilance approach are a particular focus here, as enunciated by the Harper Conservative government, with roots extending back into the Chrétien and Martin Liberal governments (Health Canada 2006, 2007).

These changes at a regulatory and related policy level are only gradually being put in place by Health Canada and its agencies. For that reason we devote more analytical attention to the broad regulatory modifications and aspirations of the system and somewhat less to product approvals, although handling the latter is central to the conceptual basis of the blueprint regulatory reform debate.

Networked power and the emergence of networked governance interests are central to our analysis of the bio-health realm. This realm is developing within a dynamic context of networked power involving competing scientific and governmental ideas, and differing institutions and interests about health policy, research, and products generally, and about bio-health products particularly.

There are interest groups and interests and there are networks, but analytically the two descriptors must be combined in the bio-health realm, because interest groups and interests are themselves networked and partnered in their actions. Also, networks are inherent in the sociology of science, in the nature of biotechnology, and in increasingly intricate ways by policy-mandated and financed granting arrangements.

These networked interests and the impacts of networked power intersect with fast-moving and complex research and developments in genomics – developments arising from the mapping of the human genome, embryonic stem-cell research, and the pressures for personalized medicine, especially in the United States but also in Canada (Boyer 2010; Harmsen, Sladek, and Orr 2006; Personalized Medicine Coalition 2009). Medicine has, of course, always been personalized in significant ways when patients are interacting with doctors or nurses about individual health conditions and medical needs. The genomics-centred concept of personalized medicine 'does not literally mean the creation of drugs and medical devices that are unique to a patient but rather the ability to classify individuals into sub-populations that differ in their susceptibility to a particular disease or their response to a specific treatment. Preventative or therapeutic interventions can then be concentrated on those who will benefit, sparing expense and side effects for those who will not' (President's Council of Advisors on Science and Technology 2008, 3).

We return to these genome-centred concepts later in the chapter, showing both the promise and uncertainties they generate and the regulatory gaps they reveal for Canada, compared to the United States, given that Canadians are seeking access to these ideas and products in the de facto North American health-care market.

Though we focus on regulatory governance, the full story of bio-health and health-research support from agencies such as the Canadian Institutes of Health Research and Genome Canada are germane as well. So also are the informational data infrastructure needs that would have to be established for these various dimensions of bio-health and more personalized medicine.

In a practical and specific manner, scientific aspects of bio-health relate to products as such. In a political and organizational manner, they deal with conceptual debates about the differences between *science-based governance* and *evidence-based governance* centred on a life-cycle approach to regulation, and with how a wider set of agencies such as patent offices, agencies for medicare funding of approved products, and agencies and interests involved in clinical trials assess evidence. Intellectual property issues, fixed on patents, figure prominently in bio-health and bring out growing concerns about patent reform centred on what constitutes an invention, and how patent-assessment criteria are applied in the bio-health realm. Related concerns are about where the lines should be drawn regarding property rights and commodification versus public goods research along with the balances of both in innovation theory and policy.

On the governance of this bio-realm, the related political and networked aspects converge initially on Health Canada, quickly expanding to other types of agencies, and extending to provincial governments where separate decisions have to be made after Health Canada's processes are completed regarding whether a province will fund new bio-health products under provincial medicare. International aspects are important too, especially in the United States, augmented by Internet-based knowledge and cross-border networks. Political advocacy and regulatory change pressures come from the larger drug industry and biotechnology business lobby, with the latter becoming rapidly a bio-health product industry rather than a bio-food industry.

Broader bio-health NGOs, patient and disease interests, and lobby groups are extremely active as organized networks in ways that are mainly supportive of greater access to bio-health products, while additionally concerned about traditional health and safety issues, and about ethics and rules regarding the conduct of research and the granting and use of patents. Despite the fact that this is a relatively new bio-realm, it has rapidly become a complex public policy community, by way of a dense set of organizational and sectoral networks with a long agenda of contending issues and perspectives.

Bio-health products cover an extensive scope of products, including biologics and genetic therapies, pharmaceutical drugs, medical devices, and radio-pharmaceuticals that contain a biotechnology-derived component. Space allows us to focus mainly on the first of these in terms of key aspects of product assessment and approval, but the broader product realms also receive necessary mention as well. We examine the broader drug approval aspects, since they, along with bio-health products such as biologics and genetic therapies, are at the core of the debate about health regulatory governance change.

Health Canada anchors the regulatory system for health products with this chapter's analysis, looking especially at its Biologics and Genetic Therapies Directorate, the department's Office of Biotechnology, and its Therapeutic Products Directorate (Health Canada 2009a, 2009b). Regulatory agencies in the bio-health realm also include the patent-approval processes and product volumes of the Canadian Intellectual Property Office (CIPO), whose work occurs prior to an application arrival at Health Canada. Indeed, many bio-health product patents never result in an application to Health Canada. The regulatory system in addition includes changing processes for pre-clinical studies that occur within private biotech companies, universities, hospitals, and other research institutions under Health Canada rules and protocols.

In the post-market aspects of the bio-health realm, the range of individual and institutional players and networks of knowledge and advice expands swiftly and comprises the medical profession, patients functioning both as individuals and through disease interest groups and networks, and government and intergovernmental agencies (federal, federal-provincial, and provincial) involved in controversial decisions on whether to fund new products for coverage under medicare and other health-care plans. Also relevant are the genomics-centred concepts regarding personalized medicine but without, at least until now, much regulatory attention in Canada, compared to the United States.

The first section of the chapter focuses on overall *regulation-making* as well as related policy developments regarding bio-health. We discuss how and why the regulatory system was designed initially and how it has been changing in a quite ambitious yet difficult manner in the last decade in particular. The second section looks at bio-health *product assessment and approvals* regarding biologics and genetic therapies. Our attention here includes some needed discussion of a genetic tests product case study of the issue already encountered in the bio-food realm: assessment of the debates and reforms regarding public versus expert involvement in decision processes on *individual products* at the patent approval stage at CIPO and at Health Canada assessment and approval stages.

The third section looks at crucial trends in *post-market regulatory monitoring* for bio-health products – the area now a major new focus under Health Canada's renewal blueprint ideas and aspirations (Health Canada 2006a, 2007a, 2009e). For our purposes, this includes aspects of regulatory monitoring once a product has been allowed on the market,

broader notions of overall pharmaco-vigilance, and, crucially, the processes for approving drugs and other bio-health products for funding
under medicare and other health-care programs (Canadian Agency for
Drugs and Technologies in Health 2007, 2008). Conclusions then follow.

Changing Bio-Health Regulation and Related Policy Development

Our examination of regulation-making and related policy about bio-
health requires discussion about (1) health regulation changes in Health
Canada, (2) patent regulation in CIPO and patent policy in Industry
Canada, and (3) specific decisions to establish new institutional entities
in biotechnology at Health Canada. Table 5.1 captures the array of laws,
policies, rules, and agencies involved in bio-health.

Table 5.1: Bio-health regulatory agencies, mandates, and regulations at a glance

Federal and related agencies and departments

- Health Products and Food Branch (HPFB) of Health Canada
- Biologics and Genetic Therapies Directorate (BGTD)
- Therapeutic Products Directorate (TPD) of HPFB
- The Marketed Health Products Directorate (MHPD)
- Public Health Agency of Canada (PHAC)
- Biotechnology Office of Health Canada
- Canadian Intellectual Property Office (CIPO)
- Commissioner of Patents
- Patented Medicine Prices Review Board
- Canadian Agency for Drugs and Technologies in Health (CADITH)
- Canadian Expert Drug Advisory Committee (CEDAC)
- Provincial health ministry drug formulary approval authorities

Mandate laws, policies, and regulations

- Canada Health Act
- Food and Drugs Act
- Food and Drugs Regulations
- Patent Act
- Financial Administration Act (Cost-Recovery Fee Regulation)
- Access to Information Act and Privacy Act
- Controlled Drugs and Substances Act
- Public Service Modernization Act
- Health Canada Blueprint for Renewal I and II (2006, 2007)
- Smart Regulation Report and Agenda

- World Trade Organization and the TRIPS Agreement
- Access to Medicines Policy

Products regulated (examples)

- Blood and blood products
- Cells, tissues, and organs
- Gene therapies
- Viral and bacterial vaccines
- Therapeutic products produced through biotechnology
- Radio-pharmaceuticals

New Health Product and Bio-Health Institutional Entities within Health Canada

Governance changes are often first made manifest by the creation or reorganization of agencies and other institutional entities. We take initial note of three: the formation of the Therapeutic Products Program in 1997, the establishment of the Office of Biotechnology in 2000, and the establishment that year of the Biologics and Genetic Therapies Directorate (BGTD).

Initial institutional change came in the formation of Health Canada's Therapeutics Products Program (TPP) (Doern 2000a). It amalgamated the former Drugs Directorate and Medical Devices Bureau of Health Canada so that it could better deal with three product groupings: pharmaceuticals, medical devices, and biologics and radio-pharmaceuticals, as well as product interactions among them. The 1997 changes stressed traditional and crucial safety mandate issues though they also introduced more explicitly goals such as timely access to such products and a more innovative regulatory system. During the 1990s, three other studies also fed directly into these changes: the Gagnon and Hearn reviews of drug and medical devices programs, respectively, together with the Krever Commission review of blood regulation. Overall, the changes began to change the TPP from a traditional health and safety-focused science-based regulator to a science-based risk-benefit manager (ibid.); themes we see below become even more broadly and openly embraced in the *Blueprint for Renewal* initiatives of 2006 and beyond.

In 2000, Health Canada established a Departmental Office of Biotechnology, which, although not itself a bio-health product regulator, was an explicit institutional recognition of the growing role of biotechnology in the health realm generally. Its main areas of responsibility are:

- to be the focal point for biotechnology in Health Canada;
- to provide sound strategic advice on the applications and impacts of biotechnology on health and lead Health Canada's horizontal policy work on biotechnology within the federal government;
- to develop, coordinate, and implement activities to build awareness, familiarity, transparency, and confidence in Health Canada's role in biotechnology and the regulatory system for biotechnology; and
- to establish and maintain partnerships with other biotechnology stakeholders, including both federal and international biotechnology regulators and associations (Health Canada 2009b, 1).

The third change to take special note of regarding governance and policy was establishment of the BGTD as a separate directorate for biologics (previously grouped under the therapeutic program of 1997), with genetic therapies also given specific mention and recognition institutionally. Thus the new directorate deals with products such as blood and blood products; cells, tissues, and organs; gene therapies; viral and bacterial vaccines; therapeutic products produced through biotechnology; and radio-pharmaceuticals.

The BGTD, as defined by Health Canada, is 'the Canadian federal authority that regulates biological drugs (products derived from living sources) and radiopharmaceuticals for human use' in the Canadian marketplace and health system (Health Canada 2009a, 1).

Biologics differ from other drugs for human use in that, in addition to the information required for other drugs, they must include more detailed chemistry and manufacturing information to ensure the purity and quality of the product.

Bio-Health Regulation-Making and Related Policy Change

As the new century proceeded, several regulation-making and related policy-change pressures occurred, three of which we highlight in Table 5.1 for their content and the political interests and values they reveal at work: the *smart regulation* agenda linked to business pressures by drug and bio-health industries and by health, patient, and disease interests to speed up the regulatory process for competitive reasons and for health product access reasons; the Health Canada *Blueprint for*

Renewal federal initiative and related consultations; and *changes and debates about patent and related intellectual property policy*, along with their crucial links to different values in public science versus property rights over health and bio-health products, and over exactly what drives innovation and in whose interests.

The *smart regulation agenda and related business and patient pressures* were an important development. The 2004 report of the External Advisory Committee on Smart Regulation (EACSR) culminated a smart regulation agenda phase about federal regulation overall, and its examination of the drug review process was one of its five case studies (EACSR 2004). The case study stressed its complex regime characteristics and agreed that safety should be the systems paramount concern, but it also argued that drug review and approval, as of 2004, still took too long, despite recent improvements. On the issue of access to new drugs by Canadians, the report highlighted the need to improve the regulatory regime to remove 'disincentives for the development of new pharmaceuticals in Canada and lost opportunities for innovation' (ibid., 79).

When examining Canada's inherent regulatory capacities compared to those of the United States, the smart regulation report argued that 'Canada cannot support a review agency as large as the Food and Drug Administration (FDA), nor can it afford to carry out drug reviews as extensive as those of its American counterpart. It must be strategic in its use of its limited resources' (EACSR 2004, 82). Furthermore, it argued that Health Canada and its regulatory partners needed to look more strategically at using international regulatory cooperation, particularly with the United States but with others as well. Health Canada has to compete at some level with the better established reputation and power of the FDA (Carpenter 2010). Given the FDA's resource endowments, expertise, and political power, this is forever a catchup game for the smaller regulator in the smaller country.

The smart regulation report recommended that Health Canada 'should focus on determining the areas of the drug approval process for which an independent approach does not contribute to the quality of the decisions or generate a benefit to Canadians' (EACSR 2004, 83). In short, it was calling for Health Canada potentially to specialize in certain areas of product regulation and accept the decisions of other national regulators (especially those of the Americans) in other realms where they had comparative advantage in expertise and overall capacity.

Coincident with the smart regulation agenda were strong pressures from the Canadian biotechnology industry, as well as the larger drug industry, about the need for regulatory speed and efficacy, particularly given the increased and high volumes of bio-health products. Previously, we emphasized the growth of Canada's biotechnology firms as mainly a bio-health industry and its fragility regarding the small size of firms, their problems of getting access to capital, and their dependence on holding often only one or two patents. This dual picture of innovative vigour and corporate vulnerability has been presented by the industry's lobby group BIOTECanada throughout the last decade. BIOTECanada highlights how its polling of public opinion shows high levels of public support for health biotechnology nationally and regionally across the country (BIOTECanada 2008). In the 2008–9 deep economic recession, the picture presented was gloomier, as the bio-health industry put the accent on the financial weakness of bio-firms that had less than one year of capital funds and on the need for bio-health to be considered urgently under infrastructure support initiatives (BIOTECanada 2009).

Of course, the bio-firms business context is closely related to the larger political economy of the Canadian and global drug industry. Globally, the situation is characterized by big-pharma firms whose patents on highly profitable, large-market, blockbuster drugs are now ending and are, therefore looking for drug inventions themselves either by doing it internally or by buying up smaller, more innovative, but capital-starved biotechnology firms (*Economist* 2007b, 2009). Canada is a part of this larger development, with its pharma political economy further characterized by international/national and regional divisions of interest between multi-national research drug firms located in Montreal and generic drug firms based mainly in Toronto.

These developments and other pressures linked to the 2009 bank crisis and deep recession, causing the drug industry globally and in Canada to look for more alliances with rivals and regulators as well to develop new products that might otherwise fail. Big treatment breakthroughs were fewer, and their costs of development were soaring (Jack 2009, 2010). These included pressures to speed up clinical trials on patients. As Jack observes, 'The more quickly that promising new medicines are launched, the greater the revenues before patents expire. By the same token, the more swiftly that drugs with problems are identified, and abandoned before expensive late-stage testing begins, the lower the wasted development costs' (Jack 2009, 7).

For these and related reasons, the bio-industry has been emphasizing the necessity for regulatory and policy change, because of the much higher volumes of smaller market bio-health products. But to begin with, there is a need to ask, is there a certain amount of hype to these predictions about future high volume growth? Some of this is simply tactical and part of a lobbying effort to get regulators to change their behaviour and get up to speed in the global marketplace. Estimates certainly vary, depending on what is being counted and what stage of research and regulation one is talking about.

The prospect of much higher volumes in bio-health products is evident and plausible. For Canada, data in 2004 already indicated that the number of products and processes in the pipeline at all stages of development had grown markedly. In 1998, fewer than 20 bio-drugs were on the market, with 6 accounting for 80 per cent of sales, and about 300 bio-therapeutics were in the pipeline. In 2004, 540 bio-drug products were at various stages of development.[1]

Health Canada data on the number of approvals of biologics from 1994 to 2009 show a significant increase in the volume of products authorized. The data cited below are based on the issuance of a Notice of Compliance (NOC). From 1994 to 1999, an average of fifty NOCs per year were granted for biologic products. These increased significantly in 2000–5 to an average of seventy-four NOCs per year, and in 2006–9 to an average of ninety-six per year.[2] To obtain a NOC, the BGTD, through the expertise of its front-line assessment teams, must determine that the benefits of the product outweigh its risks and that, whatever risks, they can be mitigated.

Given Canada's smaller market size, the 2004 data on products at all stages of development are well below data for the United States for the same period shown below. The average time of eight years for completion of the full cycle of development and regulation means there is a backlog even before one takes into account the true impacts of greater volumes and the changed technology and nature of the products. Industry interviews suggest that the knowledge base doubled in the first five years of the new century, with a significant shift to platform technologies where computer simulations are being used for new products and suites of new products. In a few cases, this has led to some firms reducing the eight-to-twelve-year clinical trial process to six to twelve months.

In the United States, there were at that time more than 370 biotechnology drug products approved and in the market, a seven-fold increase in the previous decade.[3] In 1993 seven products were approved, but in 2003 thirty-seven products were approved. From 1996 to 2003, the average number of approvals was twenty-eight – three times the average in 1986–95. Meanwhile, biotechnology *patents* granted per year in the United States were well over 7,500 per year on average for 1998–2002 – again triple the figure of the previous decade. The exact scale of predicted volume estimates varies, but no doubt the volume is growing and regulators must be able to deal with this volume nationally and internationally.

The FDA has taken steps to improve its capacity to regulate this higher volume of products in the United States and has reduced approval times needed. Given that the United States has ten times the population of Canada and ten times the market, and that the FDA is the lead global pharmaceutical regulator, it is not obvious or necessarily practical that Canada can or should reply to the volume issue by increasing its own capacities, as suggested by the smart regulation report.

The American data are of interest also as a direct indicator of product pressure on Canada. Once bio-health products are available in the United States, public and private demands grow for them to be available in Canada, from proponent companies seeking Canadian regulatory approval, from Canadian patients engaged in Internet-based self-diagnosis, and from members of the medical community. In this Internet era, with the many more patient and disease lobby groups in Canada along with many more self-informed patients in an aging but educated population, individuals readily learn about prospective new bio-health drugs, devices, and other products, and in turn they exert pressure on national regulators to approve them or to allow them to be imported.

Not surprisingly, these linked issues lead to greater advocacy and action on the need for international regulatory cooperation and harmonization. Europe, for example, in the face of bio-health product pressures in particular, established the European Medicines Evaluation Agency, now in its fifteenth year of operation. Its existence was based on policies designed to ensure that smaller and even medium-sized EU member countries could take advantage of the greater scale and capacity of scientific assessment capacity in the single European agency and not have to reproduce it themselves for their own smaller market (Abraham and Lewis 2000, 2003; Vogel 1998).

Economies of scale in U.S. bio-health regulatory capacity means that a much larger proportion of actual or potential drugs and bio-health products may not be available in a middle-sized country such as Canada, since the local market may be too small to support the cost of regulatory review. This characteristic is important in its own right, but it is ultimately tied to other previously mentioned features of market and patient pressures.

In terms of product volumes, the logic of drug development suggests that there will be even higher volume of pre-clinical studies and clinical trial–authorization processes than actual eventual new drug applications, all of which have to be assessed by firms, applicants, and Health Canada.

A second policy development during the last decade is Health Canada's *Blueprint for Renewal* and *its very ambitious and complex aspirations and arguments for change* in response to the pressures highlighted above and to other issues emphasized by Health Canada. In 2003, Industry Canada was considering a new blueprint strategy for biotechnology, given the emergence of bio-health products. No such Industry Canada–led blueprint strategy emerged for bio-health, although under the Harper Conservatives in 2006, the blueprint discourse did take hold, this time in the form of Health Canada's overall *Blueprint for Renewal*, an October 2006 discussion document (Health Canada 2006a). The Blueprint agenda was linked to changes under way in the United States that saw the most significant pressures for legislative change to the FDA since the 1960s, propelled by new needs for post-market review and new forms of evidence for such assessments (Evans 2009; Fox 2008).

The Health Canada *Blueprint* report covers both health products and food, and our summary here focuses on the drug regime aspects of the report. Health Canada's own case for renewal is a self-critical admission that its approaches over the past twenty years have not been up to speed for the world it now faces and had faced for some time. The *Blueprint* report points out that since 1953, the Food and Drugs Act and its regulations have 'largely intended to be a consumer protection statute' (Health Canada 2006a, 6). Accordingly, Health Canada's new approach to regulation identifies five challenges:

- An outdated regulatory toolkit that is increasingly limited and inflexible in responding to today's health products and food environment
- The regulatory system's current incapacity to consider a given product through its entire life cycle, from discovery through to examining the 'real world' benefits and risks of a health product or a food on the market
- The impact of social and economic changes such as accelerating scientific and technological advances, the rise of trans-border health and environmental threats, and more informed and engaged citizenry

- A regulatory system that works in isolation from the activities and policies at the stage of research and development, and those of the broader health-care system
- A regulatory system with insufficient resources for long-term efficiency and sustainability (Health Canada 2006a, 6–7).

The *Blueprint* document goes on to discuss each of these challenges and self-diagnosed inadequacies, most of which, we must add, had been a part of the larger evolving processes of much more muted criticism since at least the early 1990s.

Among its recommendations for change are suggestions for moving to:

- a product life-cycle approach, including suggestions such as the introduction of pharmaco-vigilance plans as a requirement for pre-market submission review;
- regulatory interventions proportional to risk – a process that would require the revamping of the product categorization system;
- a proactive and enabling regulatory system, so as to not only keep pace but to be ahead of the trend where possible (partly through greater regulatory foresight programs and activities regarding new and changing technologies);
- a system that makes the best use of all types of evidence, by complementing the current pre-market assessments with more extensive post-market monitoring;
- an emphasis on specific populations, regarding patient drug responses, and including a capacity to deal with new tailor-made products for diseases that affect smaller patient or genetically specific populations; and
- an integrated system (Health Canada 2006a, 7–24).

Other analyses have been critical of the innovation- and access-driven logic of the *Blueprint* report and of the previous smart regulation logic (Lemmens and Bouchard 2007). These critiques see that logic as being an excessively business-driven way of thinking rather than a public health and safety outlook, and they suggest that, by now agreeing with it, Heath Canada is weakening its core safety-focused regulatory mandate.

Following the initial announcement of the blueprint, seven regional consultation processes were conducted with a wide range of stakeholders, along with an e-consultation. These exercises produced criticisms similar to the above-mentioned general criticism about the Health Canada regulatory mandate shift, although there also were supportive comments by business, patient groups, and health-research interests. Agreeable feedback included endorsing a life-cycle approach as opposed to the current 'point in time' approach; strengthening post-market surveyance; addressing product categorization; and enhancing the transparency of decision processes (Health Canada 2007a). In the wake of these responses, a revised *Blueprint for Renewal II* was announced in 2007.

The revised blueprint began with a vision statement that gave primacy to the health and safety mandate, but its slightly revised objectives section focused overwhelmingly on the broader array of changes set out in the first blueprint document (Health Canada 2007a). Interestingly, in no part of this strategy document was the idea of the precautionary principle even mentioned. However, a broadened notion of *evidence-based* rather than just science-based decision-making was stressed. We return in the second section of the chapter to some of these crucial details of the blueprint, especially its advocacy of a progressive *licensing framework* (Health Canada 2009f), since they reveal crucial changes in bio-health governance, its complexity, and the exceptional difficulty in implementing it.

The third change to emphasize during this recent period centres on *debates about intellectual property*, chiefly regarding patents, particularly as they are influenced by innovation, high volumes of varied bio-health products, and the question of who ought to own which kinds of patented inventions. The fundamental nature of these debates needs to be understood as a feature of bio-health regulation-making. Indeed the patent process is not just a further feature; sequentially, it occurs *before* new bio-health drugs or other bio-health products reach the Health Canada assessment process for actual market approval.

Patent aspects of regulation-making and related policymaking include the following features that we discuss in this subsection: the core trade-off in basic patent policy; the strong encouragement by Industry Canada, some government labs, and funders to increase the rate of patenting as an indicator of innovation and commercialization in the knowledge-based economy; and emerging counter-pressures and arguments about the limits of patents and property rights because of the need for public science in the larger innovation process and because of the dangers of excessive patenting. These features of debate, research, and regulatory and policy change were both specific to bio-health and broader. In the next section on product approvals, we examine these features in relation to the case study discussion on the patenting of genetic diagnostic test products.

At the core of the IP-patent system is a fundamental trade-off in policy emphasis: the *protection versus dissemination trade-off* (Doern and Sharaput 2000; Drahos and Mayne 2002; Dutfield 2003). A patent ensures protection for the inventor regarding the invention and hence rewards ingenuity. Nevertheless, once a patent is registered, the patent office makes the invention public so that others may view it, study it, and potentially improve on it through a further invention. Many patents are improvements on previous patents. Once a patent is granted, it confers on the holder a monopoly property right of twenty years, provided patents remain registered through the payment of an annual fee.

Patent examiners in CIPO examine patent applications against three criteria for granting a patent. First, the invention must be new. Second, it must be useful, functional, and operative. Third, it must show inventive ingenuity and not be obvious to someone skilled in that area.

Another feature of this trade-off has been the role played in the past by requirements for *compulsory licensing*. These requirements, which now are seriously restricted, were once based on the belief that inventions made and patents granted should actually benefit consumers and society, and thus if the inventor was not going to turn the invention into a product or process actually available in the market, others should be able to, by paying a licence fee to the patent holder. Licensing certainly exists in the current system, but persons or firms that want to be able to use the patented item or process then have to negotiate with the patent holder and, in the end, the patent holder may decline to license. Licensing is at the centre of patented genetic diagnostic tests.

With respect to innovation, patents do not themselves produce innovation in the form of actual products and processes available in the marketplace. Rather, patents register and legitimate claims to possible innovation opportunities. Innovation of this more complete kind occurs only when firms or other inventors who hold patents can acquire investment capital to fund not only the needed steps to get through the rest of the subsequent regulatory approval system for products but also the manufacture, marketing, and sales stages of the full innovation and commercialization.

In this overall *innovation strategy* context, federal policy generally supports the growing importance of IP as a crucial feature of commercial originality and advance, by ensuring that Canadian companies are patenting and are aware of their patent rights and of the need to patent in the innovation age (Doern and Sharaput 2000). Important changes in Canadian policy occurred more in the 1990s, particularly after stronger patent and other IP rights emerged following the Uruguay Round of trade negotiations that led to the creation of the World Trade Organization and the TRIPS agreement (Doern 1999). In this climate of greater focus on innovation rather than just S&T or R&D strategies, federal policy fixed more on indicators such as rates of patenting rather than on earlier focal indicators such as R&D spending as a percentage of GDP. Rates of patenting in the global league tables of industrial development were seen as more *output-oriented* indicators than R&D.

Meanwhile, emerging counter-pressures and arguments about the limits of patents and property rights began to occur in response to the need for public science in the larger innovation process and because of the dangers of excessive patenting. All patents are issued by nation states, and national patent systems basically focus on the protection side of the core protection-versus-dissemination trade-off. As the Doha trade negotiations began in 2001 and as criticisms emerged since then, the focus has turned more to the dissemination side of this policy equation – to the public goods and related public science and public knowledge side of the trade-off (Dutfield 2003).

At the level of the WTO, this shift in focus to dissemination transpired through changes to the TRIPS agreement that was intended to create greater *access to medicines*, including especially bio-health products by WTO member countries with insufficient or no manufacturing capacity in the pharmaceutical sector, that would allow them to make effective use the compulsory licensing allowed under TRIPS. In Canada this change took the form of federal legislation to create Canada's Access to Medicines Regime achieved through amendments to the Patent Act (Health Canada 2006b; Mills and Weber 2006).

Under the provisions of TRIPS (Article 31), compulsory licensing or government use of patents is allowed without the authorization of the patent owner. One of the conditions under which this can be done is when such use is predominantly for the supply of the domestic market. However, TRIPS also prevents WTO members with manufacturing capacity from issuing compulsory licences 'authorizing the manufacture of lower-cost, generic versions of patented medicines for export to countries with little or no such capacity' (Health Canada 2006b, 1).

In 2003, WTO members agreed to a waive stipulation regarding this provision, whose purpose was to 'facilitate developing and least-developed countries' access to less expensive medicines needed to treat HIV/AIDS, tuberculosis, malaria and other epidemics' (Health Canada 2006b, 2). Canada was the first country to announce that it would implement this waiver. In May 2004 Canada's legislative framework was given parliamentary approval, and a year later its regulatory provisions came into force. Alas, as the result of continued industry opposition and other Canadian and international bureaucratic inertia, the policy has had little actual effect (*Economist* 2007; Mills and Weber 2006).

Both the WTO's waiver and later permanent amendment on this matter, and the Canadian regime to implement it, were motivated by wholly desirable broader foreign and international development and health policies. However, the policy of access, backed by legislative and regulatory change, additionally had to ensure that Canada was still complying with its overall international obligations under TRIPS and also the North American Free Trade Agreement (NAFTA), was respecting the integrity of its own patent law, and was responding to the competing industry and NGO interests involved in this matter.

Access to medicines is one issue in the much larger debate over the patent regime. Other analyses are specific to biotechnology and draw attention to the actual unwieldy, natural bio-world (which would not be patentable), in contrast to practices and pressures in patent law to create 'discrete immutable biological "objects" ' (Carolan 2010). Similar biotechnology-focused criticism centres on the need to rein in the scope of patents in the interests of garnering public trust and contributions to public benefits and knowledge (Caulfield 2009a, 2009b; May 2009).

Criticism relates even more broadly to concerns about excessive patenting, the ease of getting patents for questionable inventions, and the rise in some fields of patent thickets that are harmful rather than conducive to innovation (Castle 2009; Conference Board of Canada 2010a; *Economist* 2007; Mills 2010). These concerns are crucial. Their links to bio-health are further seen in bio-health product approvals and genetic testing products.

Pre-Market Assessment and Approvals of Biologics and Genetic Therapies

Since the *Blueprint* renewal is a work in progress, and because they are linked to patent issues and the particular dynamics of biologics and genetic therapies, governance dynamics in the pre-market assessment phase for bio-health products must be seen in relation to three features: (1) what the basic assessment cycle looks like and its relation to the progressive licensing framework that is a central feature of the *Blueprint for Renewal* being gradually implemented; (2) the importance of more tailored products aimed at smaller markets than normal pharmaceutical drug products; and (3) an illustrative case study about patented genetic diagnostic tests and the issues of patent, product, and health research that it raises. As is always the state of affairs, this phase of the governance analytical story is a broader drug and health products story and also a biotechnology and bio-health story.

The Assessment Cycle in Brief in Relation to Higher Volumes, More Tailored Products, and the New Progressive Licensing Framework

The full drug review under the Progressive Licensing Framework as set out by Health Canada (2009c) consists of the following stages:

1. Pre-clinical studies
2. Clinical trial authorization
3. Special access program
4. Submission review
   - Regulatory product submission
   - Safety, efficacy, and quality review
     ◦ Drug identification number application
     ◦ New drug submission
   - Market authorization decision
5. Public access
6. Post-market
   - Surveyance, inspection, and investigation
     ◦ Averse reaction
   - Post-market changes
     ◦ Drug identification number change
     ◦ Changes to a new drug

While the 'stages' portrait is a crucial one to appreciate, it is not in itself a sufficiently dynamic picture of the bio-health realm's product assessment. At the pre-market stage, the core relationships tend to be between drug firms and other research and medical applicants (and their S&T staff) and Health Canada assessors and reviewers. Consequently, these relationships are intensely bilateral and detailed.

The process begins with the complex pre-clinical studies and clinical trial authorizations, where research costs to drug firms (when trials succeed but also fail) are of increasing commercial concern, and where research ethics are also complex and often problematical.

The drug development and review involves core internal relations within drug firms and applicants, as new drug products come into view through internal self-regulation and scientific review. Thus, all of these front-line assessment features that we examined in a bio-food context apply in the faster-changing and even more complex processes of bio-health.

A further policy-related feature for biologics and genetic materials concerns the central mandate of the BGTD. In addition to the normal assessment process, manufacturers must present substantive scientific evidence of a product's safety, efficacy, and quality, of the product itself and its *manufacturing* standards/processes. This is done to ensure that a product is not contaminated by an undesired microorganism or by another biologic. If authorized for market use, biologics are then monitored in the post-market phase by being placed on a lot release schedule tailored to their potential risk, manufacturing, and testing and inspection history.

The nature of Health Canada's discussion of its new Progressive Licensing Framework is also crucial here in both a pre-market and post-market sense. The concept paper that discusses this framework is among the most revealing that Health Canada has published about its past approaches, changing challenges, and regulatory governance aspirations (Health Canada 2009f). The *progressive licensing framework* is cast as a 'life cycle and evidence-based approach,' based on good planning and accountability. Key features summarized by Health Canada are as follows:

- The central concept of Progressive Licensing is that, over time, there is a progression in knowledge about a drug. The emphasis of the new framework is to identify opportunities within this progression over the full *life cycle* of a drug, rather than placing the focus primarily upon pre-market assessment, safety, and efficacy. This represents a fundamental shift from the idea that the pre-market testing of a drug assures its safety and efficacy. The new proposed model is that a drug should be evaluated throughout its life cycle for its benefit-risk profile.

- The framework is meant to support *evidence-based* decision-making (throughout the life cycle). (This is later explained in relation to how there is a scientific standard of evidence among regulators for clinical efficacy and safety and that this standard 'requires positive validated outcomes from adequate, randomized, controlled, and confirmatory studies' [Health Canada 2009f, 8].) However, the broadened notion of evidence-based analysis arises from the fact that 'benefit-risk is founded upon scientific evidence of safety and efficacy but also encompasses a larger scope of evidence regarding contributing circumstances, including effectiveness' (ibid.).
- *Good planning* involves planning at 'every viable step in the regulatory process, which would allow for a proactive approach to managing both expected and unexpected issues' (Health Canada 2009f, 2).
- *Accountability* is associated with all aspects of the framework, including the underlying accountability of both Health Canada and drug manufacturers. These include 'the ability to make enforceable conditions upon issuing an initial market license, so that, for example, certain field reporting commitments or further studies are required to be completed' (Health Canada 2009f, 3).

We see here a compelling logic for the need for this kind of life-cycle and progressive licensing approach. It is evident that in bio-governance terms, this approach involves intricate analytical capacities of Health Canada, other public bodies, and related medical and disease networks of expertise, which they may not fully possess in these early years of regulatory regime shift.

Tailored Products and Small Markets

We have referred already to the notion of more *tailored products and small markets* in the context of the political economy of the big pharma and small bio-health firms. More specifically, the notion of tailored products and small markets refers to a shift to many more products – drugs, medical devices, and combined drugs and devices – derived from the genome mapping and DNA characteristics of small sub-populations where resulting products are directed at very small sub-markets and niche markets. Drug products now are aimed typically at huge blockbuster markets, but bio-health products are actually different (*Economist* 2007a). The extent to which this happens depends on the availability of

risk capital for firms, especially for the numerous small firms developing these products.

Similar pressures, although on a lower volume scale, were encountered in the past concerning so-called *orphan drugs*, which would not
otherwise have reached the market because of market imperatives but
were crucially needed and useful to smaller sub-markets of patients.
The United States and the European Union developed special tax-based
incentives to enable some such products to get to market (Maeder 2003),
but Canada has no such system, seemingly because it just did not register as a higher-priority tax measure. There will, however, be many more
such potential orphans in the new tailored bio-health product market
that is at hand.

One impact is that drug/bio-health regulators may have to make
greater use of their provisions for *exceptional circumstances* – regulatory
provisions that allow faster approval or temporary use, such as has
been used for some AIDS-related drugs and products. Health Canada,
as noted earlier, operates a Special Access Program, which provides
access of non-marketed drugs to practitioners treating patients with
serious or life-threatening conditions when conventional therapies
have failed, are unsuitable, are unavailable, or offer limited options.

The likely outcome of the combined growing volume and complexity and the *tailored* product dynamic is that there may be more
and more *exceptions* or 'special access' needs. With these products
approved and available in other countries, and likely flowing into
middle-sized countries such as Canada, and therefore increasing the
amounts under special access programs, it can be anticipated that it will
raise serious questions about health, safety, and benefit-risk assessment;
about is the value added to the regulatory system; and about whether
it is worth the cost in larger health-care budgetary and priority terms.

Embryonic Stem-Cell Research, Genomics,
and the Push for Personalized Medicine

This cluster of developments and issues in bio-health are of increasing
importance, revealing differences in the Canadian and United States
responses, levels of politics, governance, and controversy. We begin
with embryonic stem-cell research and then move to genomics and the
push for personalized medicine.

Embryonic stem cells are found exclusively in the early-stage
embryos, the body's master cells from which all of the body's more

than two hundred types of tissue grow (Boyer 2010; Kolata 2010). This makes them of great interest to medical researchers, both with regard to testing drugs and eventually to help repair human tissues. The central controversy about such cells and such research is that the extraction of embryonic stem cells generally requires that embryos be destroyed. Those opposed to such research on moral and religious grounds believe that such embryos have a right to life. There are some alternatives to embryonic stem cells, such as adult stem cells and reprogrammed adult skin cells, but thus far these have not developed to any great extent yet may well flourish in the coming years.

The stronger presence of religious forces in U.S. politics than in Canadian politics meant that stem-cell research was immediately and highly politicized. In 2001 President George W. Bush banned the use of federal funds for most stem-cell research, although cells created before 2001 are not banned, and private funding in the United States is allowed. President Barak Obama ordered that federal funding could support work on newer lines, and the U.S. National Institutes of Health changed its funding guidelines accordingly. In a subsequent 2010 court case, a federal judge ordered a preliminary injunction on federal funding, which was in turn overturned by a federal appeals court. The latter allowed federal funding of stem-cell research to proceed temporarily until the court rules on the merits of the Obama administration's position.

There has not yet been an equivalent level of controversy about stem-cell research in Canada. Such research is still funded, including federal funding for the Stem Cell Network, whose website headline descriptor highlights how it is 'catalysing tomorrow's breakthroughs' (Stem Cell Network 2011). The Stem Cell Network does encourage discussion about the main religious concerns about stem-cell research (Knowles 2010).

The stem-cell controversy so far is mainly a *research* controversy, quickly moving to a trials and drug product phase, where the nature of the debate will be more complex. It will effect Canada when both U.S. and Canadian product and drug applications come on o the market and into health-care systems (Baylis and Herder 2009).

The genomics-centred push for personalized medicine is a further manifestation of tailored products and small markets. But it is much more than that. As Harmesen, Sladek, and Orr (2006, 1) point out, 'Our knowledge of gene variants (person to person variation in the DNA sequence) has increased rapidly since the finalization of the sequencing of the Human Genome Project and the determination of the frequency of the DNA variation (polymorphisms) in people of different backgrounds (The HapMap project).' They go on to stress that 'this new genomics and proteomics knowledge ... will undoubtedly result in important changes in how we diagnose and treat many common and chronic diseases' (2). These likely changes in diagnoses and treatments include complex diseases such as cardiovascular diseases, obesity, diabetes, Alzheimer's, and cancer – diseases on which half of Canadian health-care expenditures are spent. These diseases are complex, because they have numerous interacting causes and effects involving multiple genes and different groups of genetic variance (Personalized Medicine Coalition 2009).

All early studies stress the promise of genomics and personalized medicine and also the huge technical, medical, ethical, and governance complexities entailed. Their full discussion is not possible here but, as with stem-cell research, some developments can be usefully highlighted in regulatory and related governance developments in the United States versus Canada, with the former moving more quickly and Canada minimalist at best.

In the United States, the FDA has recently established a post for a genomics advisor to coordinate the FDA's efforts to address the subject of genetic data and prescription drugs. A key issue here is how the FDA will evaluate genetic and bio-marker-based tests to identify patients most likely to benefit from a drug while also considering its safety for others who may receive it for other purposes (Humphries 2010). The FDA has also published guidance or draft guidance documents for voluntary pharmaco-genetic data submissions and for pharmaco-genetic and other genetic tests.

As yet, there are no obvious equivalent announced developments at Health Canada. The words *personalized medicine* do not emerge on the Health Canada regulatory or policy websites, although they are bound to be part of the background developments that are inherent in the regulatory blueprint reform changes.

An even sharper contrast is notable when American legislation is compared to its absence in Canada. In the United States, personalized medicine has received considerable legislative attention, including passage in 2008 of the Genetic Information Non-Discrimination Act (GINA). Among its chief protections is a fundamental right to privacy that 'clears the way for widespread use of genetic information in medical records and clinical decision making' (Personalized Medicine Coalition 2009, 13). U.S. President Obama has had a considerable personal interest in personalized medicine in the context of U.S. health-care reform and when, as a senator in 2006, he introduced a bill, the Genomics and Personalized Medicine Act (ibid., 14).

In Canada, there are genomic research developments at Genome Canada and in health research, federally and provincially, and related products and tests in the regulatory approval pipeline. There is, however, also a significant legislative and regulatory lag. Our earlier mention of an NDP private member's bill on protections against genetic discrimination suggests that changes may be afoot in Canada, but they are sluggish at best (Lemmens, Pullman, and Rodal 2010). For some time, criticisms have been expressed about the lack of regulation in Canada of genetic testing, including services offered by American and Canadian companies (OECD 2010; Williams-Jones 1999). The development and practice of genetic tests in Canada has taken place mainly within a clinical research and genetic counselling context, including guidelines by the Canadian College of Medical Geneticists, rather than through government-approved rules or guidelines. We return to this possible regulatory gap in our analysis of the bio-life realm.

The Need for Limits on Patents: The Patented
Diagnostic Tests Case Study Debate

Concerns have been apparent for some time about the effects of granting patent rights over human genetic materials in the health sector (Canadian Biotechnology Advisory Committee 2006; Health Canada 2007a). A particular manifestation of these concerns has arisen in relation to control over access to patented genetic diagnostic tests. Several high-profile cases triggered apprehension when patent holders exercised their rights in ways that many regarded as harmful to the provision of health services and to innovation, in that they impaired access to patented genetic diagnostic tests in areas of public health research. A related concern is the potential for patent holders to exact excessive rents by charging high prices for their products or services, thereby imposing heavy cost burdens on the health system.

One such case involved Myriad Genetics. Ontario and several other provinces objected to how the firm exercised its patent rights and called for the inclusion of a compulsory licensing provision in the Patent Act. An Ontario report argued that such an amendment 'should not obligate the provinces to first negotiate with patent holders for a license' (qtd in Canadian Biotechnology Advisory Committee 2006, 9). Section 19 of the Patent Act does not require a priori negotiation where the use is public and non-commercial.

The Canadian College of Medical Geneticists points out that the demand for these testing services is increasing 'as the number of disease genes identified and the number of tests offered in North America increases ... Genetics was once considered an esoteric speciality that did not affect the services provided in other parts of the hospital. Even five years ago, most referrals to labs came from genetics clinics and specialized doctors ... Now there is an increasing trend for referrals to come from family physicians and a wider variety of specialists' (Allingham-Hawkins 2007, 9). Because of these developments, Health Canada and Industry Canada jointly asked the Canadian Biotechnology Advisory Committee to review these biotechnology-patent regime issues. In turn, CBAC asked an Expert Working Party to examine the issue, and its report was presented to CBAC in 2005. CBAC then sought further input from stakeholders and issued its own views of the matter.

CBAC generally agreed with the Expert Working Group's main two-level approach to dealing with the problem, expressed as prevention and remediation.

- *Prevention*. Modify the patenting process to prevent patents from being issued that grant IP rights that are too broad (taking steps to reduce the potential for abuse): establish guidelines for licensing of IP rights under patent that promote behaviour consistent with the public interest and that are fair to the patent holder, and provide statutory exemption from claims of patent infringement of certain uses of patented inventions.
- *Remediation*. Provide statutory remedies for dealing with cases of abuse, and use market regulatory measures and competitive methods for increasing bargaining power to meet certain objectives (for example, moderating process of products and services) (Canadian Biotechnology Advisory Committee 2007, 2).

CBAC itself went on to comment on other key issues in the debate, including some areas where it disagreed with the expert panel. A summary captures some of the issues involved in the larger dynamics of innovation we noted earlier. CBAC concurred with the expert panel in several matters:

- Steps should be taken now to improve the patent regime and its operation in order to broaden the opportunities for mutual advantage, to deal more effectively with undesirable consequences of the exercise of patent rights when they do arise, and to improve the timeliness and transparency of patent processes.

- Human genetic materials should not be excluded from patentability on ethical grounds, and to do so 'would set Canada apart from other countries, including its major trading partners. These and related concerns can be dealt with outside the patenting process.'
- Diagnostic methods should not be excluded from patentability or providing an exemption for their clinical use, as such actions could seriously slow innovation in this field.

CBAC further agreed with the expert panel's views of sections 19 and 65 of the Patent Act. These legislative sections allow governments and other potential licensees respectively to apply to the commissioner of patents to use patented inventions without the permission of the patent holder where they have been unable to secure licences on reasonable terms. Since neither governments nor other potential licensees have apparently availed themselves of these provisions, there is no evidence that they are inadequate. Accordingly, CBAC saw no need to reintroduce a general compulsory licensing provision in the Patent Act. CBAC went on to add the caveat that other steps should be taken to enhance sections 19 and 65 of the Patent Act (Canadian Biotechnology Advisory Committee 2007, 3–4), which included developing clearer definitions of 'government use,' criteria to test the reasonableness of terms and conditions that patent holders might employ regarding licences, and other definitional issues (ibid., 3–4).

Canadian discussions of these issues relate to the work of international bodies such as the Organization for Economic Cooperation and Development (OECD). Although it is not an international regulatory body, the OECD's review of biotechnology and patent issues has focused on closely related 'soft law' instruments of guidelines and best practices. This work led to the elaboration of *Guidelines for the Licensing of Genetic Inventions* (OECD 2006), which were developed by national experts and involved consultation with a wide range of stakeholder groups across the biotechnology-patent regime. As one OECD official put it, 'Permeating the Guidelines is the idea that innovation is well served by a wide availability of technology, research and information' (Sampogna 2007, 17). As soft law, these guidelines are more exhortative than regulative. Impacts on research freedom are also a feature of the guidelines. Concerns here centre on licensing from the private to the public sector, where overly restrictive confidentiality clauses may stifle research and research diffusion, either through actual restrictions or uncertainty about such restrictions.

Finally, other issues in this debate deal with commercial development. A considerable challenge in getting a genetic product or service to market is that of negotiating multiple licences. To address this obstacle, the guidelines encourage the exploration of mechanisms such as patent pooling and standardized licensing agreements.

The ultimate front line of these combined regulatory and soft law issues regarding genetic patents is occupied by patients and medical clinicians, with the latter serving as the mediator between patients and laboratories. As McGillivray observes, 'Case studies reveal the vulnerability of patient and patient family interests in the rollout of genetic health technologies and genetic testing in particular. As with public trust in scientific research, there is a question as to whether patient, community and public trust in genetic services are being maintained' (McGillivray 2007, 44).

Health-care providers seek to provide clinically useful tests. But the Myriad case, referred to above, resulted in unequal front-line health-care delivery. In British Columbia, for example, testing was halted due to the threat of litigation by Myriad Genetics. As McGillivray further observes, 'Overnight, the price of a test became $3800 for initial cases, and, if a mutation was found, an additional $500 per family member ... This immediately separated families into haves and have-nots. Few people took Myriad's test, even though it had the advantage of producing much faster results. Testing was not resumed under the public system for two years' (ibid., 44).

Relative to a decade ago, many more voices contribute to the debate – groups that are critical of, or sceptical about, the automatic positive linkage between patenting and innovation. These voices argue that some forms of patenting are harmful to innovation and damaging to the public nature of invention and knowledge transmission in biotechnology and in other commercial fields.

Post-Market Regulatory Monitoring and Assessment for
Funding under Medicare and Health-Care Plans

The post-market phases of health-product regulation, especially regarding drugs, have always had more policy emphasis than has been the case historically for bio-food. However, the logic of the above bio-heath realm features and of Health Canada's blueprint and progressive licensing reform agenda is that the post-market is beginning to receive increased emphasis and resources. In this section, we take note briefly of two features of the post-market: one that Health Canada does include in its licensing stages map and discussion, and one that it leaves out: the longer post-market process for making decisions about which approved drug products, including many bio-health products, should be funded under health-care plans.

Basic Post-Market Rationales and Features

Health Canada's renewal discussion of post-market rationales and features is reasonably straightforward. Greater post-market activity is needed simply because drug use by diverse larger populations occurs in this much longer phase when drugs and bio-health products are in use. In addition, in bio-health areas where smaller yet more numerous sub-populations are targeted, closer post-market monitoring is required. Initially, this is also likely to increase the number of drugs that might have to be recalled or reviewed after adverse or unforeseen effects discovered through post-market assessment of actual product use, efficacy, and impacts, as discussed above regarding Health Canada's blueprint renewal plans.

At the post-market stage, the players greatly increase in number, functioning in complex networks of interests, as tens of thousands of patients, health professionals, and health institutions become a daily, even hourly, part of the post-market review, functioning as complex information and reporting networks. At this stage, the system as well becomes ever more multi-level in its regulatory governance participants, including federal, provincial, local, and international agency involvement (Doern and Reed 2000; Murphy 2006). Relationships with the FDA in the United States are also close and frequent at the technical and regulatory level of the post-market as well as at the pre-market stage.

Once a product is approved for marketing in Canada, it is monitored for product safety and effectiveness for the life cycle of the product in Canada. This follow-up role for biologics and genetic testing is conducted by BGTD in concert with other Health Canada agencies listed in Table 5.1, including the Marketed Health Products Directorate and the Public Health Agency of Canada, formed in 2003 after the severe acute respiratory syndrome or SARS crisis.

The post-market phase includes processes for reporting on adverse reactions and processes that can change the identification of drugs and changes to a new licensed drug. Much of the adverse reaction reporting has been a part of the concept of *pharmaco-vigilance*. Under the Health Canada reforms, pharmaco-vigilance is being sought even earlier in the process, through planning to detect adverse reactions earlier in patient care and possibly even in the very early non-clinical testing stage. Thus many of the planned provisions for the post-market are likely to produce more informational and data demands by applicants at the pre-market stage.

Evaluation and Decisions to Fund New
Approved Drugs under Health Care

The part of the post-market that Health Canada product regulators do not explicitly refer to in their summaries of blueprint regulatory stages concerns the evaluations and decisions to fund new approved drugs under health care. In part this omission is because other parts of Health Canada have jurisdiction here, not to mention the even more crucial role of the provinces and territories.

The ultimate drug-funding part of the post-market phase – a long-standing and central feature, of course – is receiving more resolute attention since the establishment of the Common Drug Policy in 2003 and because of the growing number of new drugs (including bio-drugs) approved to which Canadians want quick and safe access. We look briefly at this phase for bio-health products and more generally. This also has relevance for our discussion of the bio-life realm and for the expanded meanings given to *personalized medicine* and the politics it generates.

This part of post-market approvals includes the Canadian Agency for Drugs and Technologies in Health, the organization that develops and provides evidence-based clinical and pharmaco-economic reviews to assess a drug's cost effectiveness. CADTH is centrally involved in the Common Drug Review (CDR), which involves a single process to assess new drugs for potential coverage by participating federal, provincial, and territorial drug benefit plans (CADTH 2008). Provincial and territorial health ministries are a crucial part of this regime, because in the end they make the decisions (regulatory and expenditure) on whether a new drug will be funded and listed for reimbursement under medicare services delivered at the point of inpatient care.

We have also referred briefly to the *public access* stage of the drug regulatory regime and process. CADTH-linked reviews are used by the Canadian Expert Drug Advisory Committee, an independent advisory body of professionals in drug therapy and evaluation, which makes recommendations on what drugs to include in the formularies of the participating drug plans.

In more particular terms, the CDR process consists of seven basic stages:

- Complete submission received
- Submission as well as information received through independent literature review search reviewed by clinical and pharma-economic reviewers
- Reviews sent to manufacturer for comments
- Manufacturer's comments sent to reviewers for replies
- Reviews, comments, and replies sent to CEDAC and participating drug plans
- CEDAC deliberation
- CEDAC recommendation and reasons for recommendation issued to drug plans and manufacturer; final CDR reviews sent to manufacturer for information (CADTH 2007, 7)

The ultimate regulator/decision-maker regarding what drugs most Canadians have access to under medicare are the federal, provincial, and territorial governments. This includes the Quebec provincial government, though it is not a part of the CDR process because it has its own advisory committee. The provincial health ministries' formulary authorities examine the CDR recommendations and retain final say over which drugs to include in their respective formularies. In making these decisions, governments are both a regulator and spender, since, if coverage is approved, they are deciding to increase their budgets on these new drugs, and increasingly bio-drugs.

The CDR's basic roles are to conduct 'objective, rigorous reviews of the clinical and cost effectiveness of new drugs,' and to provide 'formulary listing recommendations to the publicly funded drug plans in Canada (except Quebec)' (CADTH 2006, 1). Canada's first ministers have also directed those involved in the CDR to expand its role by making recommendations 'for reimbursement to all drugs and to work towards a common national formulary' so as to eventually produce more consistent access to drugs across Canada (19). Canada's public drug plans differ, both in what drugs they cover and in the populations they serve, such as members of the military and their families, inmates in prisons, and public servants under varied federal and provincial jurisdiction. Moreover, decisions the governments need to make regarding each new drug involve questions such as how the drug compares with alternatives, which patients will benefit, and whether the drug will produce value for money.

Ultimately, these kinds of aspirations for access to funded drugs are driven by the political economy of health care (Maslove 2005). One aspect of such access is certainly the notion that national medicare is a social right or public policy entitlement. This has never been strictly true for access to funded drugs. Funded access to drugs given to patients in hospitals while under treatment are funded, but otherwise drugs are paid for or may be paid for under private plans, if Canadians have such coverage when they are not in hospital.

A second aspect of access to new drugs is the inherent political volatility if access to drugs is to be a part of what the British refer to as 'postal code' health care, where the drug access you get depends on where you live. Medicare as an equality right overall is often cast as a crucial feature of Canadian national unity, both in general terms of social citizenship and in tangible terms of intergovernmental transfers and budgetary politics (Doern, Maslove, and Prince 1988).

In contrast to these social-political aspects, a third feature of access to new drugs is found in studies showing that drug costs are the fastest-growing driver of overall health-care costs, and that health-care costs are the fastest-growing and largest part of provincial budgets in an era characterized by an aging population. A Conference Board of Canada study concludes, 'The rate of increase in drug spending has consistently outpaced the overall rate of increase in health care spending since 1984. Total Canadian spending on drugs, estimated at $18.1 billion in 2002, now accounts for 16 per cent of all spending on health care. This is up from 12 per cent in the early 1990s and 8.8 per cent in 1975' (Conference Board of Canada 2004, 33).

This means that provincial governments want to promote access to new drugs while controlling their cost and use, in part by not approving some of them in their formulary list or by delaying their listing.

It follows that the Common Drug Review can be seen as an access to drugs part of the regulatory system and also as a part of a new built-in procedural complexity. It is a reasonable procedure in the sense that decisions on inclusion in the formulary list need to undergo a form of new or further drug analysis. This further analysis potentially goes beyond what

Health Canada does, so as to deal with the further three questions posed under the CDR phase, each of which raises different kinds of procedural and value-based issues for the formulary inclusion decision-makers.

In its early years of operation, the CDR was criticized for not performing efficiently and not providing access. The industry lobby group RX&D, which represents Canada's research-based pharmaceutical companies, released an initial evaluation of the CDR in October 2005, arguing, 'The evaluation unveils exactly what patient groups and RX&D feared would happen with this added layer of review of new drug submissions which seems to function with limited or no transparency: further delays in making innovative medicines available to needing patients ... In fact by the time the CDR recommendations are reviewed by federal, provincial and territorial governments, an additional five months, if not more, has been added to the time it takes to get medicines to patients' (Canada's Research-based Pharmaceutical Companies 2005).

A further RX&D evaluation, a year later, argued that the CDR was a barrier to access. The evaluation included a comparison with patient access to new drugs in Sweden, Switzerland, France, and the United Kingdom, showing that the Canadian CDR approved only twenty-six out of fifty drugs (52 per cent), compared to Sweden (82 per cent), Switzerland (80 per cent), the United Kingdom (76 per cent), and France (58 per cent). This evaluation concluded that the CDR 'has resulted in duplication and delays in listing new medicines' (an average delay of more than seven months) and that 'the CDR process is a disincentive for more research and development into new medicines' (Canada's Research-based Pharmaceutical Companies 2006).

These analyses by industry interests present the issue entirely as a regulatory barrier, since, in their view, all the health and safety concerns had been dealt with at the earlier Health Canada pre-market review stage. A similar but more complete view is advanced in a 2007 Fraser Institute assessment (Skinner, Rovere, and Glen 2007). The study compared the global wait for the development of new drugs with Canada's, with Canada's wait times being an amalgam of three processes: for marketing approval from Health Canada; for a reimbursement recommendation by the CDR; and eligibility for provincial reimbursement. The consolidated average wait for access looked separately at pharmaceuticals and at biologics across these three elements. The Fraser Institute study concludes, 'The data for the period from 2001 to 2005 suggest that Health Canada's approval times have improved for pharmaceuticals since 2003 but have gotten longer for biologicals. Comparing the wait internationally (averaged across both biologicals and pharmaceutical medicines, and across all drug submission types) suggests that Health Canada's performance also improved relative to Europe's in 2005 but has been longer than in both the United States and Europe in the majority of the five years studied. This segment of the wait for access to new medicines equally affects all patients in Canada, whether they pay for their drugs through private insurance, out of pocket expenditure, or public drug programs' (ibid., 1–2).

After reviewing the CDR's effect on the wait time, the study reported, 'The CDR's contribution to the wait time for new pharmaceuticals is 257 days compared to 186 days for new biologics. By comparison, the wait for provincial approval of reimbursement adds, on average, another 201 days for pharmaceuticals and another 187 days for new biologics. The total average wait for approval of reimbursement is 458 days for pharmaceuticals and 373 days for biologics. In other words, patients who are dependent on public drug benefits that are delivered only through in-patient settings must wait more than a year after Health Canada has certified a new drug as safe and effective before they have access to it' (ibid., 3).

The Fraser Institute study also drew attention to the fact that new drugs covered by private insurance are generally eligible for reimbursement as soon as Health Canada has certified that they are safe and effective but that private insurance sometimes has annual coverage limits that might expose patients to significant out-of-pocket costs.

As expected, CADTH sees its overall record quite differently from these critiques by industry lobbyists. In a 2007 statement to a House of Commons Committee, CADTH stressed that, 'prior to the CDR, the (drug) reviews often took longer, and the level of rigour varied considerably across the jurisdictions. Evidence shows that the total time to formulary listing has not increased since before the inception of the CDR. This is despite establishing a standardized process that has both increased the level of rigour of the reviews and added many transparency elements to the process' (CADTH 2007, 5).

CADITH noted further that one evaluation of the CDR concluded that 'the impact of the CDR on participating drug plans has, according to drug plan representatives, been wholly positive. The drug plan's perceptions were that the process is rigorous, consistent, and has reduced duplication' (CADTH 2007, 5). Its 2007–8 annual report continued vigorously to defend its performance and record (CADTH 2008).

By 2007, CADTH had received ninety-four submissions and issued sixty-eight final recommendations, while drug plan formulary listing was recommended for about 50 per cent of drugs reviewed. CADITH also stressed that drug plans' decisions have followed CDR recommendations 90 per cent of the time. Further aspects of regime development are now underway. The CADTH budget has increased to $5.1 million to enable the CDR process to cover new indications for old drugs and for initiatives to increase transparency.

Regulatory politics and concerns about the CDR and drug access also occur at the individual level and through NGOs representing the interests of patients. For instance, the Colorectal Cancer Association of Canada has lobbied intensely for cross-Canada coverage of the colon cancer drug Avastin, which is currently covered only in British Columbia and in Newfoundland and Labrador (Laucius 2007). The drug can cost $35,000 for a course treatment. Some patients are paying for the drug themselves after failing to obtain coverage by their home province. This kind of highly personalized health-care regulatory and funding politics has garnered innovative responses from drug companies. In the United Kingdom, the drug company Janssen-Cilag has proposed a scheme whereby the U.K. National Health Service (NHS) should pay for the drug, which costs £18,000 per patient, only when the drug worked (BBC News 2007). Under the United Kingdom's approximate equivalent to Canada's CDR process, it was recommended that multiple myeloma patients showing a full or partial response to the drug after a maximum of four cycles of treatment would be kept on it, with the treatment funded by the NHS.

Conclusions

In contrast to the relative stability and entrenched nature of the bio-food realm, governing bio-health is characterized by intense change, networked power and interest governance, greater multi-agency complexity, and considerable uncertainty. As biotechnology moved strongly into high-volume research, production, and use of new bio-health products in markets and in public and quasi-market health-care systems, bio-health governance in Canada implicated a wider range of structures and agencies, including the medical and health professions and fast-forming disease networks.

This complex governance realm and its networked power characteristics are lodged within a larger health-governance domain, as well as other broader policy areas such as intellectual property and trade, with which biotechnology continuously intersects. The governance system extends – as it inevitably must, given Canada's federal system of government – into provincial government and related aspects of multi-level regulation. Genomics and personalized medicine indicate that U.S. regulatory and legislative developments loom large in a fast-changing international and immediate cross-border and Internet context.

Health Canada anchors the governance system in bio-health – a system that includes a complex bundle of policy values, product rhythms and volumes, and several agencies with diverse mandates within Health Canada. The bio-health realm extends further to players and agencies involved in patent approvals and patent policy, the funding of approved drugs under health care, and the increasingly complex post-market aspects where the medical profession, researchers, and networks of citizens as patients and carers are intricately involved every day.

The bio-health realm raises particular issues about patents, patent assessment processes, and overall intellectual property policy and regulatory affairs. Mounting stress on property rights and the protection of bio-health product patents have been challenged on exceedingly valid and important grounds about the need for a renewed emphasis on public goods, and public knowledge in the closely linked policy and governance realms of biotechnology, health care overall, and the broader dynamics of what it takes to produce economic and social innovation.

The bio-health realm is governed by a much more positive and accommodating politics than in the bio-food realm. While there are health risks associated with bio-health products such as biologics and genetic testing, the overall pattern of interest group support is positive. The great majority of interests are seeking greater and faster access to these products. Concerns about genetic privacy and discrimination have emerged, and the absence of both rules and laws in Canada are a primary governance gap.

The big-pharma industry and the smaller but dynamic biotechnology firms operating in Canada and globally seek efficient but also efficacious assessment and approval processes at the patent regulator, CIPO, at Health Canada, and eventually via the CADTH process. Network-based patient groups and disease groups are anxious for access to new products, including those that may already have been approved in other countries, especially in the United States. Research interests and networks are largely positive although also concerned about newer and more complex risks and benefits that accompany the products themselves and the ethics of the research process. We have seen this in the Canada-U.S. comparison regarding the debate about embryonic stem-cell research.

While the chapter has covered all three aspects of regulatory governance – regulation and related policymaking, pre-market product approval processes, and the post-market – the last of these has received our closest attention. The central reason for this is Health Canada's *Blueprint for Renewal* strategy with its expressed policy and governance philosophy. The *Blueprint* report covers health products and food but we have emphasized the drug regime. Health Canada's own case for renewal is broadly a self-critical admission that its approaches over the past twenty years have not kept up with the world it now faces and has faced for some time.

We maintain it is essential to understand the main features of the blueprint and current efforts to implement it in both governance and networked interest-based democratic terms. The life-cycle approach, the progressive licensing framework, evidence-based versus science-based governance, and pharmaco-vigilance all stretch and extend the bio-health (and health) governance system in time spectrums covered, spatially across the country, and among interests and stakeholders required to cooperate more with each other while anchoring their own core interests. These governance changes have major implications for the capacities needed to undertake this complex work: scientific and technical personnel, increased levels of funding in a recessionary budget era, high levels of trust, and accountability systems that are not at all easy to make or keep transparent.

On product approval, higher volumes are indeed a key feature of the bio-health regime. So also are the more highly tailored, small markets and personalized health nature of bio-health products, and in biologics assessment; so is the need to assess manufacturing processes and standards as well. Post-market regulatory governance lies at the heart of what an extended regulatory life cycle really means. What is more, post-market regulatory governance is politically significant because of the increasingly delicate choices, and assessments of choice, regarding which new products will be eventually approved, and if so, how quickly, and funded by the provinces under health care.

6 The Bio-Life Realm: Playing Catchup with Self-Disciplined Power Relations

Introduction

Bio-life is the most recent biotech governance realm and certainly remains an emergent field in Canada's biotechnology system of governance. We look here at the origins and evolution of the regulation-making features of the bio-life realm through developments and debates about new reproductive technologies in the late 1980s and early 1990s, with further references to developments following the announced mapping of the human genome in 2000, and events that are more recent. The bio-life realm exhibits the importance of self-disciplined power relations and changed complex kinds of self-regulation.

Reproductive technology was the focus of the work from 1989 to 1993 of the Royal Commission on New Reproductive Technologies (RCNRT) (1993; Miller Chenier 1994) and was brought to the national agenda mainly by women's groups and coalitions. There was a decade-long delay as legislative and regulatory action took place starting in 1996, finally resulting in legislation in 2004 and subsequent regulations up to 2007.

The commission's final report referred only briefly to biotechnologies in a chapter on commercial interests and identified at that time very few Canadian biotechnology companies engaged in research directly relevant to reproductive technologies, with most research taking place in universities using funding from government or private foundations (RCNRT 1993, chap. 24). Nonetheless, the reproductive technologies debate, led by women, noticeably included early aspects of genetic testing and presaged many of the larger dimensions of the later and current forging of the bio-life realm of biotech governance in the genomics and even wider bio-life era (Abraham 2012).

While the commission was aware that Canada had joined the Human Genome Project in 1992, the commission did not view the project goal of determining the structure and location of human genes as part of its mandate with respect to genetic research and technology as related to human reproduction. However, other aspects of the bio-life realm distinctly do centre on the burst of product and related activity following the mapping of the human genome in 2000 and the resultant push for genomics, reflected in new versions of more personalized medicine, and in the formation of Genome Canada.

Groups with a direct interest in assisted human reproduction, genetic research and its applications, and genomics include organizations representing women, children, families, industry/business, persons with disability, faith and religion, individuals and families, insurance and law firms, medicine and clinical laboratories, scientific research (natural and social sciences), universities, and governments. All these are bio-life networked interests within Canada's liberal democracy and plural society.

Both the federal and provincial levels of government intervene in the role of medicine and other professions, and their relationships with the public, through policies on which biotech products are permitted or prohibited, on reporting obligations for practitioners using such regulated products, and on the organizational location for activities and practitioners. These interventions demonstrate what in earlier chapters we called institutionalized power: governing through state structures and processes.

However, medical power – that is, the influence and authority of not just doctors and related professions but also the drug industry, health sciences, and therapeutic professionals within society – stems from a number of sources and relationships, including a body of expertise based on systematic knowledge, skills, and training developed though evidence; relatively high societal prestige and public legitimacy; and substantial occupational autonomy through delegated authority from the state.

Professional power also operates through self-disciplined forms of power, upon which we focus particularly in the bio-life realm. This occurs when a patient (or the patient and significant other and perhaps other family members) internalizes and acts upon advice from medical specialists and other practitioners alongside influences from personal and group identities. This form of power 'reaches into the very grain of individuals, touches their bodies, and inserts itself into their very actions and attitudes, their discourses, learning processes, and everyday lives' (Foucault 1980, 39).

Women were a particularly crucial interest in advocating the need for the Royal Commission on New Reproductive Technologies and in its subsequent research, findings, and recommendations. In 1987, feminist organizations, researchers, and health groups, who argued that discussions on reproductive technologies were adjudicated primarily by the medical profession with little attention to the broader implications of these technologies on women, initiated the Canadian Coalition for a Royal Commission on New Reproductive Technologies. Pressure from both the medical/scientific community and women's organizations found space inside and outside the royal commission, and in the commission's final report.

As in the bio-food and bio-health realms, questions arise and debates ensue in bio-life policy over what ought to be regulated, what should be supported, and who is an expert on what matters. Who is a valid representative of what groups and concerns? And what counts as knowledge when making decisions and regulating behaviours? Perhaps more so than in the companion bio-realms, in the bio-life realm politics and policy, ethics and moral considerations feature alongside science, technology, and intellectual property.

In the bio-life realm, the notion of what is a *product* takes on meanings that lose many of their traditional comfortable moorings compared to those of the other two bio-realms examined previously. While both bio-food and bio-health products can be assessed at the pre-market and post-market phases of assessment and regulation, and even withdrawn from the market or the health and the food system, bio-life aspects cannot be recalled in anything like the same sense. The defective 'product' may be a child or offspring for whom the pre-life or early stages are crucial and need to be effective in profoundly human terms.

Since the concept of 'the bio-life realm' is not a standard categorization in either public administration or science policy, we first offer some context-setting and definitional comments on this realm and then relate it to associated concepts of bio-power and bio-politics. The second section of the chapter then examines overall regulation-making in the bio-life realm and how and why the regulatory system was designed the way it is and how and why it was forged, albeit in quite tentative and slowly developing ways.

The third section focuses on pre-market bio-life product assessment and approvals, while the fourth section looks at trends in post-market regulatory monitoring and compliance processes for bio-life products and practices. In both of these sections, the above noted blurring and contentiousness of what a product is, is paramount. We use the field of assisted human reproduction as a case study to illustrate the hesitant and incomplete nature of the governance of this realm, including its statutory basis under federal criminal law, but its extensive reliance on guidelines produced by government and complex self-regulation by involved participants.

The fifth section examines key political issues regarding implementation of the bio-life realm. It illustrates gaps in the bio-life governance realm that reflect and repeat conflict and uncertainty over what the economic and social boundaries are and what state intervention means in the bio-life realm. It includes further discussion of how genomics and personalized medicine and genetic testing products are regulated or not regulated. We then look more closely at implementation and power relations surrounding the protracted and unfinished realization of the Assisted Human Reproduction Act and Assisted Human Reproduction Canada. Conclusions then follow about the emergent nature and ongoing evolution of this mesmerizing realm of biotech governance. The bio-life realm is an exemplar of regulation as networked public-private-personal governance.

Bio-Civic Regulation, Bio-Power, and Bio-Politics

Bio-life regulation is an example of 'civic regulation,' a subtype of civic regulation sometimes called moral and sexual regulation (Prince 1999). This entails rule-making by the state and, at times, by societal institutions, on numerous aspects of human behaviour and needs as well as ethical issues, moral transactions, and intimate conduct (Cruikshank 1999; Hauskeller 2007; Maheu and Macdonald 2010; Midgley 2010). Bio-life regulation comprises rules regarding egg, sperm, and embryo transplantation, therapeutic and human cloning, genetic research, sex selection, abortion, assisted human reproduction, genetic manipulations and applications, xeno-transplants, and embryonic stem-cell research, genomics, and personalized medicine (Hochedlinger 2010; Hughes 2009). It deals with the role of the state on the subject of human life itself; of people as prospective parents, children, family members, donors, recipients, patients and clients. In this regard, it involves the interplay of law, morality, self-determination, equality, and identity (Einseidel and Timmermans 2005).

A significant amount of bio-life regulation lies outside the formal structures and activities of the state. It pertains to the particular lives, experiences, and biographies of people vis-à-vis family trees, ethnic and racial communities, sexual orientations, parental aspirations, physical and mental disabilities, employment opportunities, and other personal and social identifications. As a form of civic regulation, bio-life rule-making helps to define and shape the meaning of health and well-being, procreation and childbirth, human bodies, normalcy and abnormality, and family planning, among other issues.

Bio-life regulation is not just the governmental control of domestic life, as feminist analyses have rightly noted of social welfare programs, but involves the organization of human life itself within the domestic sphere of social relationships (Newman and White 2006). Perhaps more so than for the bio-food and bio-health realms, the bio-life realm is underpinned by profoundly complex and compelling systems of moral values, religious beliefs, and ethnic influences.

The bio-life realm, therefore, includes 'reprogenetics' (Raz 2009), the application of genetic technologies to human procreation, such as pre-implantation genetic diagnosis, prenatal genetic diagnosis, in vitro fertilization, and other reproductive technologies. Genetic testing and screening raise a host of actual and potential controversial issues: genetic testing as a basis for prenatal detection of inherited diseases; determination of fetal sex for non-medical reasons; assessment of a person's eligibility or lack of it for health and life insurance coverage or for employment; the privacy and uses of genetic information; and genetic intervention and what it might mean for the inclusion or exclusion of persons with disabilities and thus the diversity of human conditions in society (Buchanan 1996; Shakespeare 1998, 2005). Bio-life regulation takes place in, and helps to shape, a moral economy of science and enabling technologies in addition to personal identities of self and social identities of families and communities. It involves the allotment and definition of gender roles, body images, and social statuses in the populace at large.

Bio-life policy and governance illustrate what Michael Foucault and others call 'bio-power' (Foucault 2008; Shiva and Moser 1995), a term that applies to the bio-food and bio-health realms as well, because they entail state measures directed at the protection, promotion, and enhancement of the life of a population in a political community. Bio-power is a positive or productive exercise of state power that not only improves individual human well-being and the health of the body politic, but generates new forms of knowledge and ways of understanding and managing human life. Other writers have called this bio-policy field of new genetics 'regulated eugenics' (Tyler 2010). In the case of bio-life policy, exercising bio-power concerns the regulation of reproduction and heredity, and the bio-medical construction of human bodies.

In turn, the notion of bio-politics as it relates to the bio-life realm refers to legal, scientific, economic, moral, and social practices and techniques through which populations are controlled and fashioned (Tyler 2010). It is not only the politics of governments or state actors. Interests and institutions converge and collide, expressed through political and philosophical standpoints, over cloning, stem-cell research, human embryos, patents and conceptions of nature, human life, and the human body (Scala 2007). Multiple perspectives and voices are at play, too, concerning the meaning of sickness, disease, impairments, the normal and the pathological, in which 'the deepest personal and collective experiences of embodiment, vulnerability, power, and morality interact with an intensity' (Haraway 1989, 4).

Bio-Life Realm Regulation and Rule-Making

Bio-life regulation-making involves the politics and processes of establishing the overall rule-making system, including its statutory mandates and its core delegated regulations, guidelines, policy statements, and industry standards. Table 6.1 shows the strategic statutory and regulatory features established for the bio-life realm at the level of the federal government in Canada.

One of the distinguishing characteristics of the bio-life realm compared to the bio-food and bio-health realms is the greater prevalence of guidelines as a governing tool, especially by Health Canada, along with professional codes and industry standards. Guidelines produced by government, and the guidance documents that accompany them, are directed at several groups and serve somewhat distinctive functions: symbolic assurance to political leaders and an interested public that action is being taken on an issue of significance; administrative support to departmental or agency officials charged with putting a policy into practice; strategic signals to industry associations and firms on what compliance with laws and rules may and may not require; and, in the bio-policy field, concrete advice to health scientists and medical professionals on the conduct of their practices and research.

Table 6.1: Bio-life regulatory agencies, mandates, and regulations and guidelines at a glance

---

Federal agencies and departments

- Assisted Human Reproduction Canada
- Canadian Institutes of Health Research
- Genome Canada
- Health Canada
- Industry Canada
- Justice Canada
- Natural Science and Engineering Research Council
- Privacy Commissioner of Canada
- Social Sciences and Humanities Research Council

Mandate laws, regulations, policy statements, and guidelines

- Assisted Human Reproduction Act, 2004; and Assisted Human Reproduction (Section 8 Consent) Regulations, 2007
- Canadian Institutes of Health Research Act, 2000
- CIHR, SSHRC, NSERC, Tri-Council Policy Statement: 'Ethical Conduct for Research Involving Humans,' 1998
- Canadian Institutes of Health Research, 'Updated Guidelines for Human Pluripotent Stem Cell Research,' 2007
- Food and Drugs Act, 1985; and Processing and Distribution of Semen for Assisted Conception Regulations, 1996; Medical Devices Regulations, 1998; Safety of Human Cells, Tissues and Organs for Transplantation Regulations, 2007
- Hazardous Products Act, 1985; and Controlled Products Regulations, 1988
- Health Canada, 1998, 'Guidance for the Risk-Based Classification System of In Vitro Diagnostic Devices'
- Health Canada, 1998, 'Preparation of a Pre-Market Review Document for Class III and Class IV Device Licence Applications'
- Health Canada, 1999, 'Preparation of an Application for Investigational Testing: In Vitro Diagnostic Device'
- Health Canada, 2001, 'Selected Legal Issues in Genetic Testing: Guidance from Human Rights'
- Health Canada, 2003, 'Preparation of New Drug Submissions in the CTD Format'
- Health Canada, 2003, 'Guidance for Clinical Trial Sponsors: Clinical Trial Applications'
- Health Canada, 2005, 'Management of Drug Submissions Guidance Document'
- Health Canada, 2006, Clinical Trials Manual
- Health Canada, 2007, 'Submission of Pharmacogenomic Information: A Guidance Document'

Products, procedures, and practices regulated or prohibited (examples)

- Tissues for transplantation
- Egg donation
- Embryo freezing and transfer
- In vitro fertilization and other infertility treatments

- Pre-natal diagnosis
- Processing and distribution of semen for assisted reproduction
- Stem-cell research
- Human cells for transplantation and reproduction
- Therapeutic cloning
- Human cloning
- Genetic testing
- Surrogacy

Understood as a relatively softer form of regulation, guidelines typically lack the force of law and statutory importance. Rather, they are administrative tools as well as processes for managing relationships between government officials and actors in the economy and civil society. Flexibility is chosen over legitimate coercive power, although as we see further below, core federal legislation anchored in the use of federal criminal law powers has faced constitutional challenge in the courts. As such, guidelines may involve two-way exchanges of information and interpretations and can entail negotiations over the principles and practices of rule-making and implementation.

The policy and governance of Canada's bio-life realm built gradually over the past twenty-five years can be described in terms of two phases of development and delay, with a third phase just beginning.[1] In the first phase, spanning the mid-1980s through the 1990s, new regulations dealing with aspects of bio-life were created through existing statutes – the Food and Drugs Act as well as aspects of the Hazardous Products Act – connected with the mandate of Health Canada. These initiatives demonstrate the practice, later expressed in the federal regulatory framework for biotechnology (Industry Canada 1998), of using functioning laws and government departments to avoid the proliferation of new rules and agencies, and the duplication of efforts.

Nonetheless, in the second phase of development of the bio-life realm, since the late 1990s, new legislation and new agencies have been created. Legislative and agency developments in this more recent phase include the Assisted Human Reproduction Act; the Canadian Institutes of Health Research Act and its associated network of virtual research centres; and the formation of Genome Canada. These statutory and structural innovations indicate the push and pull between interests tending to support biotechnology promotion and open access (scientific and medical communities, infertile families, and certain industries), and other interests calling for strong government regulation applying a precautionary approach based on scientific evidence of benefits and harms and, in some aspects of bio-life, the outright prohibition of activities deemed harmful to society (feminists and women's groups, disability organizations, and religious and faith-based groups).

With the December 2010 Supreme Court of Canada reference opinion on the Assisted Human Reproduction Act, in which the legislative authority of Parliament to act in this area has both been constrained and confirmed, a third phase of policy development is emerging (Reference re Assisted Human Reproduction Act, 2010 SCC 61). Among other issues, this new phase suggests a limited federal role in the direct regulation of controlled activities by health practitioners and in medical facilities. And it may not be the last time the courts are called upon to interpret the legislation or to act as the umpire in Canada's federal system of divided jurisdictions. That said, the federal government could explore other political means to work through constitutional questions over the distribution of legislative powers and regulatory activities in this important area of the bio-life realm. We may see renewed efforts by Ottawa to negotiate bilateral agreements with one or more provinces on assisted human reproduction. We have more to say about this reference opinion by the Supreme Court later in the chapter.

Reflecting the growing politicization of reproductive technologies and genetic engineering, these more recent enhancements to the rule-making and funding raise the profile of bio-life policy and governance within the federal government and across the country.

At the centre of the rule-making of the bio-life realm are Assisted Human Reproduction Canada (AHRC) and Health Canada, while Genome Canada and the CIHR serve important research funding roles along with the older federal research funding councils of SSHRC and NSERC. But the above-noted federal granting bodies also have regulatory roles in research ethics (Doern and Stoney 2009). In tandem with these government actions, new interest groups and coalitions that are focused on bio-life policies and politics have emerged. The impact of moral and sexual regulations, which bio-life policies are of a kind, is often on very specific groups rather than the general public – couples dealing with infertility – generating and reflecting specific interest structures and politics (Abraham 2012; Prince 1999).

Governing instruments available to federal authorities include statutes, legislative regulations, guidelines, policy statements, and standards. As might be expected in a public policy system addressing fundamental matters of human health and well-being, statutes and regulations are a significant component of this realm. Possibly more significant and interesting is the relatively greater deployment in the bio-life realm of 'softer law' in the form of guidelines and policy statements, issued by federal departments and research councils, as actual or quasi rule-setting tools. From the perspective of liberal democratic politics and social learning, guidelines and policy statements can be seen as measures to inform and raise public awareness, offer reassurance on the efficacy of – and contribute to building a social consensus on – research activities and medical interventions in the bio-life realm. Over time, from this perspective, guidelines and policy statements may forge a level of public support that can lead to governments then introducing legally enforceable standards through regulations and legislation.

Of the institutions involved in regulating bio-life, several lie outside the Canadian state, such as families, religious organizations, and disability groups. In fact, these social structures can be thought of as both self-regulating institutions and institutions regulated by formal and informal rules generated by other structures. Of particular note is the role of the Canadian Standards Association (CSA), a not-for-profit organization, that develops national standards for a range of activities, including under the Food and Drugs Act. For example, the CSA has crafted standards for medical devices quality-management systems, ocular tissues for transplantation, lympho-hematopoietic cells for transplantation, and permitted organs for transplantation.

Other non-governmental bodies, in particular health professional associations and some industry associations, play roles in the formulation and implementation of rules governing bio-life. Examples include the Canadian Life and Health Insurance Association, Society of Obstetricians and Gynaecologists of Canada, Canadian Paediatric Society, Institut national de santé publique du Quebec, and the Canadian College of Medical Geneticists.

Beyond such forms of professional self-regulation, an even wider view of regulation in the bio-life realm, in terms of rule-making, enforcement, and compliance, involves the personal self-regulation of individuals within families, religious communities, ethnic or racial groups, and other relevant social categories. At play here are rules of behaviour backed not by the sanctions of the state but rather by social structures, cultural mores, belief systems rooted in religious faith and public morals, as well as personal dreams and fears of health risks in having a child created as a result of fertility drugs, from donated semen, eggs, or embryos, and/or because of inherited genetic disorders.

The effect of these rules and related norms and values on the considerations and choices of individuals or couples can be quite influential. A cultural responsibility or a personal sense of obligation, for instance, expressed and reinforced by experts, elders, or confidants may well outweigh the impact of a statutory right or public service on the actual decisions and actions of people. As Raz (2009, 609) elaborates, 'Individual attitudes and behaviours towards reprogenetics should thus be considered in the context of community norms and values which directly influence marriage patterns, family planning, perceptions of selective abortion, and power relations of gender, age and status.'

Accordingly, techniques of socially based self-regulation in the bio-life realm embrace a multitude of activities: gathering specialized information from experts and increasingly from the web and social media; consulting with individuals and organizations, including self-help or support groups; calculating the risks and benefits of particular technologies or tests; reflecting on core norms and internalized principles of one's sense of identity; and acting upon oneself.

A feature of this societal context of self-regulation entails what Novas and Rose (2000) call the 'genetic network,' which an individual (couple or family) may become part of through 'groups, associations, communities of those similarly at risk; groups of patients at particular hospitals or clinics; participants in trials of new therapies; subjects of documentaries and dramas on radio, television and the movies' (490). When this occurs, although never entirely detached from public rules, self-regulation involves the exercise of self-disciplined power and hence de facto regulatory power beyond the state.

Regulating Bio-Life Products and Activities: The Case of Assisted Human Reproduction

A look at the regulation of assisted human reproduction technology within the federal government offers important insights on the role of policy and governance in the bio-life realm. Stages of the regulatory process we consider here are policy and standard setting, product assessment and related inspection and enforcement, and post-market monitoring of products and risk communication.

Our survey reveals that regulating bio-life activities and products in assisted human reproduction is, to a large degree, an unfinished work in governance and policy implementation and remains a contentious and contested realm in Canadian society and federalism. In part, this is due to the long gestation period from 1989 to the present before laws, rules, and agencies were established, some key aspects of which are still in contention constitutionally in Canada's federal system and indeed resulted in 2012 in the elimination of AHRC.

Policy and Standard Setting

In federal documents and on official websites, AHRC is described as the federal regulatory agency responsible for matters relating to assisted human reproduction. This is a fair description, particularly of the *official mandate* of AHRC found in its enabling legislation. The principles underpinning the mandate, as recognized and declared by the Canadian Parliament, are that:

a. the health and well-being of children born through the application of assisted human reproductive technologies must be given priority in all decisions respecting their use;
b. the benefits of assisted human reproductive technologies and related research for individuals, for families and for society in general can be most effectively secured by taking appropriate measures for the protection and promotion of human health, safety, dignity, and rights in the use of these technologies and in related research;
c. while all persons are affected by these technologies, women more than men are directly and significantly affected by their application, and the health and well-being of women must be protected in the application of these technologies;
d. the principle of free and informed consent must be promoted and applied as a fundamental condition of the use of human reproductive technologies;
e. persons who seek to undergo assisted reproduction procedures must not be discriminated against, including on the basis of their sexual orientation or marital status;
f. trade in the reproductive capabilities of women and men and the exploitation of children, women, and men for commercial ends raise health and ethical concerns that justify their prohibition; and
g. human individuality and diversity, and the integrity of the human genome, must be preserved and protected. (AHR Act, sec. 4)

These numerous principles show how the act is 'a discursive site' (Deckha 2009, 23) of cultural, commercial, religious, scientific, gender, and legal narratives and claims. The principles furthermore show the politically negotiated and multi-layered nature of bio-life policy in prescribing standards (Caulfield and Bubela 2007; Morris 2007; Scala 2007). These principles were enshrined into the legislation after the House of Commons Health Committee was asked in May 2001 by the minister of health to consider a document titled *Proposals for Legislation Governing Assisted Human Reproduction*. The principles replaced a pre-amble in the draft proposal and became a framework used by the com-mittee to assess legislative options. As the chair of the committee noted, the Standing Committee thus became 'a forum for public consultation, a filter for contentious debate, and an agent for policy development on an issue that had already consumed more than a decade of debate' (Brown, Miller Chenier, and Norris 2003). This direct effort to engage Canadians through their elected federal representatives followed a 1995 voluntary moratorium on certain related practices such as cloning of human embryos, buying and selling of eggs, sperm, and embryos, and other activities deemed to be unethical and socially unacceptable, and an unsuccessful effort in 1996 to establish legislation that would establish boundaries around such practices. In October 2002, compre-hensive legislation to prohibit certain activities and to regulate oth-ers with respect to assisted human reproduction and related research began the parliamentary legislative cycle. Reintroduced in February 2004 after prorogation, it contained multiple amendments but finally received royal assent in March 2004 (Hebert 2004).

When the 2002 legislation moved through the House of Commons in October 2003, it was by a vote of 149–109, reflecting a Parliament divided along partisan and philosophical lines. Implicit in the act's list of guiding principles is the notion of precautionary governance, mean-ing that an extensive range of cherished values must be recognized and defended while regulating reproductive technologies and genetic engi-neering (Scala 2007). Although bio-life policy in Canada has emerged within 'a pre-existing context oriented toward the commercial appli-cations of biotechnology' (Sharaput 2002, 158), the commercialization of trade in human reproductive capabilities is explicitly banned and a host of socio-ethical criteria are emphasized. Expressed another way, we can see here a form of social value–based policymaking alongside science-based policymaking – in short, ethics as evidence.

In effect, AHRC's official mandate expresses four broad objectives: (1) protecting and promoting the health of donors, patients, and off-spring; (2) protecting and promoting the human dignity and human

rights of Canadians who use or are born of assisted human reproduction technologies; (3) fostering ethical principles in research and clinical
practices; and, (4) allowing scientific advances that benefit Canadians
(AHR Act, sec. 24). Functional areas of responsibilities and program
activities set out in the legislation are as follows:

- Implementing and administering the licensing framework for
  controlled activities, including assisted human reproduction procedures and related research
- Developing an inspection system to ensure compliance with the
  Assisted Human Reproduction Act and regulations
- Developing and maintaining a national personal health information
  registry on assisted human reproduction as part of a more comprehensive assisted human reproduction health surveillance strategy
- Providing advice to the minister of health and to the AHRC Board
  of Directors
- Communicating with and engaging stakeholders on assisted
  human reproduction technology issues
- Collecting and disseminating public information

The actual *working mandate* of the agency, however, is not this extensive in practical scope or regulatory in nature because the implementation of the Assisted Human Reproduction Act is occurring in
a series of measured and cautious steps – a policy style and political
stance of proceeding carefully (Baylis and Herder 2009; Jones and
Salter 2009; Manseau 2003; Royal Commission on New Reproductive
Technologies 1993). When the legislation was enacted in March 2004,
various sections of the statute came into force in April 2004, further
sections in January 2006, and another section in December 2007. The
agency itself was not established until 2006, and the president was
appointed in 2007. The agency reports to Parliament through the minister of health, and it is the minister and Health Canada officials who
are responsible for the development of strategic policy in the legislation and regulations.

That public policy is both government actions and inactions is clearly
demonstrated in how assisted human reproduction policy and regulation is being implemented. Large portions of the Assisted Human
Reproduction Act still are not in effect at the time of writing (in early
2012), most significantly those dealing with privacy and access to information (including a personal health information registry), licensing

and administration, inspection, and enforcement. In other terms, the full regulatory policy toolkit is yet to be activated and implemented.

In consequence, AHRC's mandate in practical terms emphasizes knowledge functions: advice to the minister, communications with stakeholders, and information to the public. The agency has a memorandum of understanding with Health Canada that agrees that the Health Products and Food Branch Inspectorate of that department will undertake inspection and compliance as required, until the relevant sections of the Assisted Human Reproduction Act that deal with inspection and enforcement come into force.

The piecemeal and staged implementation of the legislation on assisted human reproduction may be due to a number of influences: the institutional and administrative inertia of large bureaucracies, the influence of social conservatism in the Harper Cabinet and government, and the national style of regulation in Canada, which we have previously dubbed 'regulatory gradualism' (Doern et al. 1999, 402).

In this specific case, another factor is particularly significant. Multilevel governance is a constitutional reality in the Canadian federation, marked at times by controversy. The bio-life policy realm, more specifically the Assisted Human Reproduction Act itself, is under such intergovernmental challenge. More than other bio-policy spheres, the cross-jurisdictional nature of the bio-life area of reproductive technologies slowed movement from the beginning and continues to render it a contested area in Quebec. Work within the federal bureaucracy to facilitate coordination and consensus at the intergovernmental level moved from efforts to develop a national framework from 1993 to 1996, through attempts to build a national consensus from 1997 to 2000, then to working through the federal Parliament in an effort to secure national action (Miller Chenier 2002).

Health-care policy and governance in Canada is a jurisdiction shared between the federal and provincial orders of government. Provincial jurisdiction over health-care services is dominant constitutionally to the extent that provinces have explicit grants of authority for hospitals and related care institutions; property and civil rights, including mental health matters and the regulation of health professions and practices; and local or private matters, including community health and municipal health boards (Jackman 2000).

Federal jurisdiction in the area of health derives from five constitutional sources: most generally, perhaps, intervention is through the federal spending power as applied to health transfers; for specific groups where it has designated responsibility, including the armed forces, RCMP, First Nations and Inuit peoples, immigrants and refugees, inmates in federal penitentiaries, and veterans; for emergency or national health matters, the peace, order, and good government power may apply; for 'patents of invention and discovery,' Parliament has jurisdiction (section 91 [22] of the Constitution Act, 1867); and for food and drugs, hazardous products, and aspects of environmental and reproductive health, it is through the federal criminal law power (section 91 [27] of the Constitution Act, 1867). According to constitutional law professor Martha Jackman, 'Parliament has relied primarily on its spending and criminal law powers as a basis of federal action in relation to health' (2000, 96).

In the overall bio-governance regime, much of the federal support for research and promotion, as through the Canadian Institutes for Health Research, for example, rest upon the spending power for their activation. And in the bio-life realm, it is the criminal law powers, rather than the peace, order, and good government powers (the latter of which the Royal Commission on New Reproductive Technologies [1993] favoured to justify federal intervention in this policy area) through which the federal government established the Assisted Human Reproduction Act of 2004.[2] Health Canada cautiously explains the legal and policy circumstances:

> On December 15, 2004, the Government of Quebec filed a reference with the Quebec Court of Appeal, challenging the constitutional validity of some sections of the AHR Act. The reference was heard in the Quebec Court of Appeal on September 17–19, 2007. The Court rendered its opinion on June 19, 2008, that all challenged provisions are unconstitutional because they are outside federal powers. A constitutional reference seeks an advisory opinion of the court. The AHR Act cannot be struck down based on an advisory opinion.
>
> After considering the Quebec Court of Appeal opinion, the Government of Canada has brought an appeal before the Supreme Court of Canada to address questions regarding the constitutionality of the Act. In the meantime, the Act remains in effect. (Health Canada 2010b)

The challenged sections are 8–19, 40–53, 60, 61, and 68 of the Assisted Human Reproduction Act, and the Quebec Court of Appeal concluded that all of these sections were ultra vires or outside the jurisdiction of the Parliament of Canada. These sections deal with licensing-controlled human reproductive activities, creating a registry of health records, and enumerating powers of inspection and enforcement and for offences and punishments. A year after the Quebec Court of Appeal rendered its opinion, Quebec's National Assembly passed the Act respecting Clinical and Research Activities relating to Assisted Procreation (R.S.Q., c. A-5.01). So while business interests and ideas of liberalized trade may favour the amalgamation and streamlining of levels of regulation, governmental interests and ideas of federalism, especially in Quebec-Canada relations, appear to favour the apportionment and division of regulatory powers.

Specifically with respect to the challenged sections of this act, the constitutional issue concerns whether the prohibitions, regulations, and penal sanctions in the legislation are suitably related to the criminal law powers of Parliament or fall under provincial powers. In a complex and split 5–4 decision, the Canadian Supreme Court held that the federal government's appeal of the Quebec court's decision be allowed *in part* (Reference re Assisted Human Reproduction Act, 2010 SCC 61). One majority of justices held that sections 8, 9, 12, 19, and 60 of the act are constitutional, meaning they are within the scope of federal criminal law; and, moreover, that sections 40(1), (6), and (7), 41–3, 44(1) and (4), 44–53, 61, and 68 are constitutional to the extent that they relate to inspection and enforcement and offences in relation to constitutionally valid provisions.[3]

A different majority – that is, a separate configuration of five justices – held that sections 10, 11, 13, 14–18, 40(2), (3), (3.1), (4), and (5), and sections 44(2) and (3) exceed the legislative authority of the Parliament of Canada, representing an attempt to regulate hospitals, clinics and medical research, property and civil rights, and matters of a local and private nature, all of which belong to the jurisdiction of provinces.

A core issue dividing the Supreme Court concerned the purpose and nature of the legislation: is the Assisted Human Reproduction Act of 2004 primarily a law to prohibit certain inappropriate activities, or does it include objectives and policy instruments aimed at promoting and controlling certain beneficial practices of assisted reproduction? In short, is it about 'the woah' of biotechnology, a criminal law with prohibitions and penalties; or is it a health law to promote the positive 'wow' of reproductive practices and products? In their reasons for questioning the constitutionality of the sections in dispute, Justices LeBel, Deschamps, Abella, and Rothstein noted,

Reproductive technologies do not constitute a matter over which Parliament or the provinces can claim exclusive jurisdiction. Two very different aspects of genetic manipulation have been combined in a single piece of legislation. The social and ethical concerns underlying these two aspects appear to be distinct and in some cases even divergent. While the prohibited activities are deemed to be reprehensible, the controlled activities are considered to be legitimate. Parliament has therefore made a specious attempt to exercise its criminal law power by merely juxtaposing provisions falling within provincial jurisdiction with others that in fact relate to the criminal power. (Reference re Assisted Human Reproduction Act, 2010 SCC 61, para. 278)

In the end, a majority of the justices saw that the act had sought to address both these aspects, hence the constitutional predicament.[4]

Interestingly, certain sections of the act in effect as of April 2004 allow – in the governance of the agency and the legislation – for a person nominated by the deputy ministers of health in the provinces to attend and participate at meetings of the board of directors of the AHRC (sec. 28); for the formation of advisory panels with outside members (sec. 33); and for the negotiation of an 'equivalency agreement' between the federal government and any province, for five years, which would recognize that laws of the province in force are equivalent to relevant sections of the legislation and corresponding regulatory provisions (sec. 68 and 69).

These elements of the legislation reflect ongoing efforts by the federal government to move to a national framework based on a national consensus leading to national action and, in the period leading up to the 2004 legislation, showed no success in the initiation of equivalency agreements, the creation of advisory groups, or the engagement of deputy ministers (Miller Chenier 2002).

Inspection, Enforcement, and Post-Monitoring Bio-Life Activities

Canada's lead agency for regulating assisted human reproduction and aspects of genetic research and applications confronts a paradox of power. Despite a preference for regulatory gradualism and the relatively greater use of softer tools for compliance and enforcement, AHRC's core mandate, in order to ensure its constitutionality, rests on the unmistakable federal powers of criminal law. This reliance upon criminal sanctions has prompted strong criticisms from within the medical and scientific communities with concerns over criminalizing medical practices and discouraging health research (Deckha 2009;

Scala 2007), as well as within the legal and bio-ethical communities, including a debate over using criminal sanctions in the moral status of embryos and the health and safety of women (Caulfield and Bubela 2007; Morris 2007; Mykitiuk, Nisker, and Bluhm 2007).

Sections of the Assisted Human Reproduction Act that are in force and not under constitutional challenge set out a range of activities prohibited in Canada. In section 5, these prohibited activities are as follows:

a. create a human clone by using any technique or transplant a human clone into a human being or into any non-human life form or artificial device;
b. create an in vitro embryo for any purpose other than creating a human being or improving or providing instruction in assisted reproduction procedures;
c. for the purpose of creating a human being, create an embryo from a cell or part of a cell taken from an embryo or foetus or transplant an embryo so created into a human being;
d. maintain an embryo outside the body of a female person after the fourteenth day of its development following fertilization or creation, excluding any time during which its development has been suspended;
e. for the purpose of creating a human being, perform any procedure or provide, prescribe, or administer anything that would ensure or increase the probability that an embryo will be of a particular sex, or that would identify the sex of an in vitro embryo, except to prevent, diagnose, or treat a sex-linked disorder or disease;
f. alter the genome of a cell of a human being or in vitro embryo such that the alteration is capable of being transmitted to descendants;
g. transplant a sperm, ovum, embryo, or foetus of a non-human life form into a human being;
h. for the purpose of creating a human being, make use of any human reproductive material or an in vitro embryo that is or was transplanted into a non-human life form;
i. create a chimera, or transplant a chimera into either a human being or a non-human life form; or
j. create a hybrid for the purpose of reproduction, or transplant a hybrid into either a human being or a non-human life form.

One unique feature of this act is its effort to 'hold onto humanity' by drawing boundaries between human beings and non-human life forms because of concerns over the 'animalization' of people. The Assisted Human Reproduction Act, as Deckha explains, 'is distinctive for its intention to define and demarcate species boundaries around the human and the animal. Unlike other legislation that merely facilitates animal use (such as hunting legislation) or even tries to codify animal welfare (such as anti-cruelty statutes), the AHRA concerns itself with boundary questions of what is acceptable for human bodies (and thus animal ones) and is constitutive or violative of human identity' (Deckha 2009, 39).

In addition to these prohibitions on human cloning, sex selection, and transplantation are restrictions on commercial activities and the commodification of procreation. 'Commodification anxiety arises from the notion that not everything can or should be bought or sold in the marketplace. For example there is widespread concern that love, babies, and sex should be excluded from the domain of the market' (Deckha 2009, 30). Numerous women's organizations and feminist researchers in Canada have long voiced such concerns over reproductive technologies (Montpetit, Scala, and Fortier 2004). Under the act, no person shall offer to do or to advertise the doing of anything in relation to these forbidden activities. Conversely, no person shall pay or offer to pay consideration to any persons for doing any activity with the purpose of creating a human being, prohibited under this law. Moreover, the act prohibits the practice of advertising for sale, and the selling or purchasing of sperm, ova, in vitro embryo, or a human cell or gene with the intention of creating a human being or of making it available for that purpose (sec. 7). These activities are to be decent gifts without a price tag (Yee 2009).

The act expressly prohibits the practice of treating surrogacy arrangements as a commercial transaction. Section 6 states,

1. No person shall pay consideration to a female person to be a surrogate mother, offer to pay such consideration or advertise that it will be paid.
2. No person shall accept consideration for arranging for the services of a surrogate mother, offer to make such an arrangement for consideration, or advertise the arranging of such services.
3. No person shall pay consideration to another person to arrange for the services of a surrogate mother, offer to pay such consideration or advertise the payment of it.
4. No person shall counsel or induce a female person to become a surrogate mother, or perform any medical procedure to assist a female

person to become a surrogate mother, knowing or having reason to believe that the female person is less than 21 years of age.

In recognition of the division of powers in Canada's federation, with health and family law primarily matters of the provinces, the act states that the validity of any agreement under which a person agrees to be a surrogate mother, made under provincial law, is unaffected by this section of the act.

Fundamental rule-making powers of the agency not yet in effect bear directly on shaping the agency's practical mandate for inspection, monitoring, and compliance. These potential powers would allow the agency to exercise authority in relation to licences under the act; collect, analyse, and manage health-reporting information relating to controlled activities; designate inspectors and analysts for the enforcement of the legislation and regulations; and issue licences for controlled activities, including clinical trials (AHR Act, sec. 24 [a], [e], [g]; sec. 41). Of the statutory powers that *are in force*, the agency may:

- provide advice to the Minister on assisted human reproduction and other matters to which this Act applies;
- monitor and evaluate developments within Canada and internationally in assisted human reproduction and other matters to which this Act applies;
- consult persons and organizations within Canada and internationally;
- provide information to the public and to the professions respecting assisted human reproduction and other matters to which this Act applies, and their regulation under this Act, and respecting risk factors associated with infertility. (AHR Act 2004, sec. 24 [b], [c], [d], [f])

What, then, are the de facto inspection, enforcement, and compliance activities of AHRC? Notwithstanding the criminal power base for the agency, a traditional command and control style – whereby government inspectors routinely exercise strong public authority over private actors – does not predominate. However, the Supreme Court's 2010 reference opinion indicates that powers of agency federal inspectors (sections 45–53) are valid to the extent that they are based clearly upon criminal law powers, suggesting perhaps more of a policing role for inspectors. At the time of its elimination in 2012, the agency had an annual budget of approximately $11 million and a planned staff capacity of about forty-four full-time equivalents.

With no inspectorate branch in place, it continued to rely on Health Canada for this important function. According to AHRC and Health Canada – the portfolio that has responsibility for developing policy and regulations on assisted human reproduction and related genetic applications – the agency's compliance strategy is to encourage and facilitate observance with the legislation and associated regulations. Described as using 'a cooperative approach,' AHRC worked with groups of professionals and non-governmental organizations concerned with reproductive bio-medicine, fertility clinics, genetic research centres, and laboratories.

This work involved promoting the principles of the legislation, distributing and sharing information on matters of human reproduction addressed by the act and regulations, evaluating complaints related to allegations of infringements of the act, and working with stakeholders (counsellors, embryologists, ethicists, laboratory technicians, lawyers, medical doctors, nurses, psychologists, scientists, and social workers) to advance compliance with the legislation and regulatory requirements. At the same time, professional associations and non-governmental organizations, such as the Royal College of Physicians and Surgeons of Canada and the Canadian Fertility and Andrology Society, engage with AHRC and Health Canada in advising on the development of policy and regulations for this area of the bio-life realm.

'Meanwhile,' as Baylis and Herder (2009, 117) observe, 'human embryo research that is not prohibited in legislation continues in Canada in accordance with existing research guidelines (*TCPS* and the *Guidelines for Stem Cell Research*) pursuant to the grandfathering clause in the AHR Act.' The grandfathering clause stipulates that 'a person who undertakes a controlled activity at least once during the period of one year preceding the coming into force of those sections may subsequently, without a licence, undertake the controlled activity for that purpose until a day fixed by the regulations' (AHR Act 2004, 71).

If the agency 'raises the possibility of non-compliance, it is the responsibility of the facility to take timely and appropriate action to comply with legislative and regulatory requirements.' The agency 'will clarify what is required to achieve compliance, but will not dictate how it must be achieved.' As a central feature of its compliance strategy, the agency anticipates 'a graduated series of interventions and measures that take into account different factors, such as the seriousness of the breach. For example, in some cases the Agency might provide the facility with guidelines to assist in compliance with the regulations' (Health Canada 2010, 3).

This does not mean, however, that strong regulatory powers are absent from the assisted human reproduction realm. If AHRC becomes aware, through reliable information, that a person or a clinic may have committed an offence involving a prohibited activity, then agency staff will refer the matter to the RCMP for their consideration. If a complaint results in the RCMP launching an investigation, AHRC will inform the person making the complaint about the outcome of the investigation and the action taken (Health Canada 2010b).

The Supreme Court's reference opinion on the Assisted Human Reproduction Act holds that a personal health information registry (sections 14 to 18) exceeds the jurisdiction of the Parliament of Canada. The effect of this opinion would seem to deny the agency an important tool for the post-market monitoring of assisted human reproduction products and practices, including those involving genetics. Such a national registry, now unlikely in the near future, could have provided invaluable services to three target audiences. First, for the general public and the assisted human reproduction policy community, information is required on bio-life issues and activities and for professionals on consent to use. Second, for prospective assisted human reproduction users, a registry could offer information on the results of research and provide reports on the results on success rates by clinics across the country. Third, a personal health information registry could provide a 'look-back' and 'trace-back' mechanism for donors, patients, and offspring born of assisted human reproduction procedures – donor-conceived people and parents of donor-conceived people.

Denied constitutional authority to develop a public registry and related knowledge activities, Health Canada now needs to reassess how it can make a contribution to science-based information and to risk-oriented communication, reporting on the safety, quality, and efficacy of treatments and products available in Canada. Timely reporting of these kinds of knowledge can assist individuals and families in making more fully informed decisions on the role of biotechnologies in human procreation, personal identity, and family formation.

A related point on the makeup of regulation, illustrated by the case on assisted human reproduction policy, is that, as a policy instrument, the legislation and rules of AHRC are not separate from the macro-political context of the Canadian constitution, federalism, and the dynamics of intergovernmental relations. AHRC's emergent compliance strategy that emphasizes cooperative actions is understandable and well-founded as a feature of *political risk management* as well as one of health and science-based risk management, but the agency's demise means that Health Canada will have to think again.

Outstanding Issues: Genetic Testing, AHRC Inertia, and the Politics of Implementation

Finally, we need to draw attention to two outstanding issues in the bio-life realm: genetic testing products and regulatory gaps and continuing inertia regarding AHRC – both of which generate further politics in the implementation of bio-life governance.

First, there are closely related and growing regulatory issues regarding fast-developing genome and genetic-testing products. These products raise questions about where the post-market starts, which regulators loom largest, and with what kinds of monitoring and review powers. There are issues regarding policy and rules on bioethics at various stages in research, product development, and review. In both areas, developments in the United States are especially germane, in part because they will have cross-border impacts and demonstration effects on Canada.

On products as such, one need only cite new genetic testing sales and marketing strategies. For example, the American company Easy DNA already markets its genetic tests in Canada and prominently emphasizes that it is ISO 17025 accredited (Easy DNA 2010). Another U.S. company, Pathway Genomics, has begun selling its genetic testing kit in over 6,600 U.S. drugstores (Stein 2010) rather than just on the Internet. These new products have not been regulated by the FDA, but they raise many concerns about their efficacy, accuracy, and use. The company's view is that it does not need to be regulated by the FDA although the FDA has faulted, albeit cautiously, these companies for selling unapproved genetic tests (Pollack 2010c). It is highly likely that that these kinds of U.S. products can and will be sold in Canada, and thus there are regulatory gaps potentially on both sides of the border, depending on when and if Health Canada moves into this aspect of bio-life regulation.

On the broader bioethics front, another U.S. development is of interest. President Obama has established a new Bioethics Commission. In contrast to the previous Bush Administration's Council on Bioethics, which had a strong theoretical and ideological emphasis, the Obama-era Bioethics Commission will focus on a range of problems that are more practical, as well as their solutions (Meslin 2010).

Canada has had no equivalent major bioethics arena of advice, policy, and debate.

Genome Canada has fostered some bioethics debate and discussion through its grants and its GELS activities, but these are quite weak and marginalized. Indeed, paradoxically, the main previous process for debating some major bio-ethics issues was the Royal Commission on New Reproductive Technologies from 1989 to 1994.

These kinds of bio-ethics issues have specific meaning locally. A compelling example here is the front-page coverage given to an announcement in June 2010 by Canadian, British, and American scientists about their research on autism and genetics (Connor 2010a). Their studies of the genomes of nearly a thousand individuals with autism compared their DNA against that of over one thousand unaffected persons. The researchers, as a result, believe that the genetic make-up of children with autism is important in causing the illness. With more research, these scientists believe that early diagnostic tests and treatment can be devised.

Bio-life technologies are especially susceptible to both the 'wow' and the 'woah' reactions and instincts. In this case, the 'wow' factor came in the announcement itself and press coverage of it. An immediate letter to the editor of the *Globe and Mail* newspaper expressed the 'woah' reaction in intensely human ways. The letter came from Estee Klar, who signed her letter as the founder of the Autism Acceptance Project (Klar 2010). Her letter expressed the views of the mother of an autistic child reacting to the announcement about genetics and autism. She wrote of the 'lack of public awareness about autistic pride and the many autistic adults who have helped our understanding of what it means to be "different."' She expressed deep concern about language that might cast persons with autism as having fouled-up genes. Klar concluded by stating that autistic people are 'more than their genes' and that like non-autistic people, they are 'shaped by [their] environment, supportive families, good education and so forth' (Klar 2010).

Of course this example deals with one disease and one announced genomic/genetic research study, although it is by no means atypical in its core bio-life features. This unfolding story combines emergent science, immediate media hype, latent hopes, and manifest fears about human consequences. These features are likely to be the norm for some time, given that a decade after the announcement of the mapping of the genome, scientific gains may be genuine and even exciting, but medical cures and gains lag far behind (Wade 2010a, 2010b). Moreover, these features and human stories reveal regulatory gaps that might or might not be filled in the near term or longer term, given their complexity.

On a broader international scale, a recent task force convened by the International Society for Stem Cell Research, with Canadian involvement via the Ottawa-based Stem Cell Network, drew critically needed attention to the array of new medical clinics established around the world (Taylor 2010). These clinics are offering patients cures for various illnesses based on the most recent advances in stem-cell research. At least some of these clinics are making false claims and using dubious ways to attract patients through Internet sales and marketing. The task force report urged new rules that would require clinics to obtain approval by national oversight regulators, and such clinics would be required to have independent ethics boards to review their procedures (ibid.). A further obvious reality emerges from these examples drawn from American, Canadian, and broader global sources: in order to be effective, regulatory actions and governance arrangements will have to become increasingly transnational in nature and coordinated in practice.

Turning to the demise of AHRC, a dense array of issues have yet to be dealt with satisfactorily. Delays in implementing one of the legislation's substantive sections leaves several pressing matters unresolved and thus generate unintended and adverse consequences for individuals and families and for fertility clinics and health professionals. Unsettled issues include immediate questions and impending items of concern about the anonymity of egg, sperm, or embryo donors, and whether adult offspring of fertility treatment should have the right to know the identity of the donor; the rise in multiple births and the associated concerns of serious health effects for these infants due to infertility drugs, in vitro fertilization, and the practice of implanting two or more embryos to increase the likelihood of a pregnancy; the underground economy in trade for human eggs and sperm, embryo, or surrogate motherhood; the long-term effects of fertility drug treatments on women who undertake these procedures, and the lack of long-term monitoring of such treatments; genetic testing of eggs, semen, or donors to pre-select certain genetic traits and to avoid other traits, including sex; what is legally allowed or prohibited in reimbursing women for surrogate-mother services and egg donors for expenses or lost earnings from work interruption (Blackwell 2009, 2010b; Carlson 2010).

These unresolved issues and unregulated activities are a source of ongoing frustration and apprehension among childless couples and prospective parents, as well as medical practitioners and fertility specialists, infertility advocates, bio-ethicists, and lawyers specializing in health and family law. From the uncertainty of the rules and what is permitted regarding assisted human reproduction, critics identify outcomes that involve inconsistent practices among fertility clinics; rules being flouted in some circumstances; couples seeking to buy embryos, eggs, or sperm from donors, and women offering surrogate services for substantial payments; and wide-ranging uncertainty over the rights of donors, carriers, and surrogate mothers in matters of family law (Abraham 2012; Blackwell 2010a, 2010b).

We therefore have a state of affairs in Canada, as one news report observes, 'where those couples quietly compensate donors for their gametes [human eggs or sperm], despite the legislation that criminalizes doing so. Where lesbian couples lie to doctors about their sexual orientation to avoid paying to quarantine a friend's sperm for six months. And where doctors and counsellors sometimes adopt the credo of "Don't ask, don't tell."' In addition, 'Donor banks used to pay men for their sperm. But since the prohibition was implemented, sperm- and egg-donation programs must rely on the altruism of anonymous donors. With some donor programs dried up, prospective parents import sperm from the United States or travel there to pay as much as $30,000 for an egg cycle. And then there are those who cannot afford to play by the rules, and who instead turn to donors they find online and pay under the table' (Carlson 2010, 1–2).

These experiences and dilemmas strikingly illustrate local implementation of assisted human reproduction policy. Health Canada is dealing with a legislative field of action contested constitutionally, and that is contentious politically with the public and within the Harper government. Yet any symbolic assurance offered by the legislation and by the soft approach to compliance pursued by the agency goes only so far. Effective implementation requires the translation of legislative provisions into rules and guidelines, as well as the commitment of adequate administrative capacities that, in turn, enable actions in accordance with the principles of the legislation. Expressed in these terms, implementation might be taken simply to be a top-down approach of putting policy into practice. A more suitable image of implementation of the Assisted

Human Reproduction Act is of a bumpy and meandering route of inaction and action, rather than a smooth and straightforward course of accomplishments.

Given the Supreme Court of Canada's multifaceted opinion on the jurisdictional status of sections and subsections of the agency's legislation, a top-down approach to rule-making and enforcement is significantly constrained in structuring this important component of the bio-life realm. The implementation for assisted human reproduction is largely being determined locally and, in certain aspects, by the actions of provincial governments. For example, as of June 2010, the Quebec government extended medical coverage for fertility treatments to women, thus making such treatments free to families – the first jurisdiction in North America to do so. At a cost of $32 million in the first year, Quebec's health ministry estimates that the annual number of embryo implantations will expand significantly over the coming years (Ravensbergen 2010).

In looking at the implementation of the Assisted Human Reproduction Act, we witness all three forms of power in operation – institutionalized state power; networked power of this policy community that includes pharmaceutical companies and professional associations and health practitioners; and the self-disciplined power of governance of the self by self-conduct – although in ways and means not always intended or expected.

On institutionalized state power, the main theme in this case is the underdeveloped exercise of the federal legislation and minimal rule-making and enforcement by the agency, which was a likely factor in the resignation of four senior staff members in 2009 and of two directors of the board in 2010, raising a red flag over the interests represented and not represented on the agency's board (Blackwell 2010b). Of course, delays in regulatory governance by the agency can be attributed, in large part, to the uncertain constitutional status of much of the legislation and to the apparent tardiness of the Supreme Court of Canada in rendering an opinion on this issue.

On the networked power characteristics of this policy community, a main theme is the co-regulation or, in this case to date, mostly professional self-regulation by obstetricians and gynaecologists, medical geneticists and genetic counsellors, family physicians and nurses, and other health professionals (Knoppers and Isasi 2004). That the different forms of power are interconnected is illustrated by actors in the policy network, such as physicians in fertility clinics complaining about the limited exercise of state power, specifically the limited guidance and enforcement by the AHRC and the lack of leadership by the federal government (Blackwell 2009, 2010a, 2010b).

On self-disciplined power, which we have defined as governance of the self by the self as embodied actors, a number of themes emerge from our analysis of the implementation of the assisted human reproduction policy. One concerns some people engaged in buying or selling human eggs and sperm, thus commercializing reproduction, contrary to the explicit purposes of the Assisted Human Reproduction Act. Another theme of self-disciplined power relates to claims by known donors to be involved in the upbringing of their offspring and claims by some children to know their genetic parents. Still another theme involves lesbian women consulting local community-based groups about becoming a biological mother and planning for parenthood (Carlson 2010). The common thread here is the importance of choices and decisions by individuals, couples, or triads, health professionals, and others in closely observing or deliberately ignoring or cleverly interpreting rules and procedures. Self-disciplined power does not equate to full compliance with state legislative policies; the discipline involved in the conduct of oneself may be aimed at realizing personal interests more than advancing public policy objectives. The case of assisted human reproduction policy shows that implementing rules depends upon the actions of several state structures, across levels of government, and of numerous societal organizations, economic and professional interests, family, and other intimate arrangements.

Conclusions

The nascent but not entirely new field of the bio-life realm first emerged in the name of concerns about reproductive technologies, and medical and health practices regarding them, where the role of women's groups have been pivotal. Then later, the bio-life realm developed under the specific impetus of the mapping of the human genome and resultant new and contested products linked to genetic testing. The bio-life realm exhibits the importance of self-disciplined power relations and changed complex kinds of self-regulation. The nature of this realm is such that rule-making is an intertwined cultural, moral, and social phenomenon as well as a market enterprise, political issue, and scientific occurrence. The notion of what a product is or ought to be enters more uncertain and disputed territory and governance space.

This bio-realm concerns the politics of human life itself. On the issue of commercialization of gestational capacities and body cells and tissue, the Assisted Human Reproduction Act declares 'that human donors are not to be paid and that a corresponding market should not merge in human reproductive materials' (Deckha 2009, 42). Related issues deal with understandings of human subjectivity and spirit, multiple discourses on human dignity and rights, and ultimately on what it means to be a person.

On the regulatory origins of this realm, we noted that its development shifted from an early reliance on existing laws and government departments to the creation of new agencies and statutory mandates to address issues of bio-life. Health Canada is the core federal regulatory institution in the bio-life realm, linked closely to the establishment of Assisted Human Reproduction Canada. Industry Canada and CIPO, with their responsibility for patents and intellectual property, are involved in bio-life governance as a direct extension of linked bio-health matters. On the research funding side, Genome Canada, the Canadian Institutes for Health Research, and the Networks of Centres of Excellence are central institutions for basic research and knowledge mobilization and for the regulation of research ethics.

A number of related overall conclusions can be drawn on what regulatory governance looks like in this realm of biotech governance and policy. One, regarding democracy, is that all five notions are apparent in the bio-life realm. In terms of representative democracy, Cabinet-parliamentary government responded to and struggled with the assisted human reproduction issues for more than a decade, culminating in a majoritarian decision in the passage of legislation. In terms of federalized democracy, the division of powers between the two orders of government conditioned the initial policy design of this regulatory regime and the delayed gradual pace of reform, and, more recently, generated the constitutional challenge to fundamental parts of the Assisted Human Reproduction Act.

In terms of interest group democracy, professional bodies in the medical and scientific communities are significant partners with government in implementing bio-life policies and practices and are substantial players in promoting and protecting their interests and autonomy for professional self-regulation.

In terms of civil society democracy, consultations have provided spaces for social-movement organizations representing relatively marginalized groups to interact with state authorities and engage in policymaking. At the same time, however, these processes and interactions can expose divisions within social movements – such as the divergent views in the women's movement among feminist researchers, infertile women, lesbian couples, single women, women in ethnic and racial groups, and women with disabilities (and *multiple intersections* of these identities) – on the merits of minimal state regulation versus the use of criminal sanctions. There are comparable divisions in the disability movement among other segments of civil society.

In terms of direct democracy, individual citizens participate in biotech governance by seeking information on reproductive products or genetics, participating in and interacting with genetic communities, making choices about personalized medicine, deciding to make or to obtain an egg or semen donation, intermingling church and state in personal discourse and action about reproduction policy, and making decisions on who will be born, determining identity-based life choices with implications for human diversity and genetics.

Given this complex interplay of democratic politics, Canadian federal authorities have adopted a cautious and gradual approach to rule-making in the bio-life realm. This incremental approach is evidenced by the length of time (fifteen years or more) to establish a law on assisted human reproduction and then an agency to implement that statutory mandate, and it is apparent in the relative absence of 'hard laws' and the greater reliance on 'soft rules' and voluntary compliance through the use of policy statements and guidelines on bio-life policy. Fukuyama (2002, 200) adds, 'The regulatory regime for human biotechnology is much less developed than for agricultural biotechnology, largely because the genetic modification of human beings has not yet arrived as it has for plants and animals.' Moreover, in Canada in particular, issues of the human condition – dealing with morality, the integrity of life, individuality, and interdependence – are inevitably debated and taken up through our federal condition of divided constitutional powers, and competing territorial interests and social identifications.

A second general conclusion points to the network-like character of this realm of biotechnology. Networks of non-governmental associations and interest groups serve a number of roles, including knowledge production and evaluation, the promotion of specialized products and therapies, and rule-making and implementation. Medical and other health professions constitute an important arena of non-state actors for bio-life regulation, information, advice giving, and interventions. The result of this dense community of interests and institutions is a fluid, multi-level regulatory governance, with the concurrent realities of practical complexities, diverse and often divergent messages, inherent tensions, and abiding anxieties over the future of human nature.

A third conclusion, which appears to follow from the first and second, is that the bio-life realm reflects new governance approaches of our times – that is, of governing, despite efforts to use criminal law powers, with relatively little or apparently less government. In a post-regulatory state, according to this perspective, governments rely on softer policy tools and on partnerships through policy networks rather than depend upon traditional command and control tools of power and resources exercised through state bureaucracies. There is, in this observation, an element of truth, although only up to a point, especially given the controversies over the federal use of criminal law powers to anchor federal jurisdiction. Most contemporary concepts of governance and 'new public management' do not capture the extensive range of different types of actors and social structures involved in bio-life policy and practice.

In contrast to many other public policy fields, the governance approach in the bio-life realm is not a case of government retreating from previous roles or responsibilities or sharing those roles with private firms and civil society organizations. It is, more accurately, a case of legislators and governments in Canada moving cautiously and reluctantly at times into regulatory responsibilities on sexuality, morality, and human nature. In the areas of human reproductive technologies and genetic technologies, and personalized medicine, the role of government in Canada and in other countries was generally minimal before the 1980s, relying on self-regulation by the medical and scientific communities.

In regulating bio-life, a number of factors present challenges. First, the deeply felt and often contrary views and religious convictions held by Canadians on family, gender, disability, human dignity, and the morality of inclusion make this a politically intricate and socially charged regulatory domain. Second, experts, whether in the scientific, medical, or ethical fields, are divided on the efficacy and desirability of various technologies and therapies, and the research on the risks and benefits of certain procedures is contested, with charges of eugenics never far from the discourse. Furthermore, policy debates are infused with strongly entrenched positions and often diverse legal principles, self-understandings, and economic interests. Third, there is the arduous challenge, underscored by the Supreme Court reference case, in drafting legislation or regulations in this realm of 'transcribing

philosophical or moral objectives into precise and operative legal ter-
minology' (Campbell and Pal 1989, 112).

The regulatory regime for bio-life always entails a delicate and contentious dynamic among governmental rule-making, professional
health practices, scientific breakthroughs, commercial developments,
individual and family preferences and expectations, and community
norms and values. In this milieu, it seems more fitting to speak of civic
regulation and the regulatory craft today and into the foreseeable
future, than of a hard and fixed regulatory regime for the political control of biotechnology.

Our fourth conclusion is that self-regulation is a critical dimension of
the bio-life realm, and that the concept of self-regulation has a meaning not often recognized or adequately examined in the public policy
or regulation literature. The usual interpretation, including in writings
on biotechnology, is that of professional or sectoral self-regulation, the
exercise of independent control by industry or the academy or scientific
community, with little direct administration or supervision by government. The notion of personal or social self-regulation, however, emphasizes the role of citizens governing their own conduct within a particular
socio-cultural environment in historical time and a political space.

The concept of 'genetic communities' (Novas and Rose 2000) underscores the existence and significance of these societal circumstances
on the choices perceived to be available and the decisions eventually
made by individuals and families. Some authors see the technological transformations of our embodied selves, through biotechnology,
as the creation of 'biological citizenship' (Hughes 2009), while others
write of human cyborgs in a post-human age of life and politics, with
the 'cyborg citizen' as 'a self-regulating organism that combines natural
and artificial together in one system' (Gray 2002, 2). In a similar fashion,
Fukuyama (2002) has argued that contemporary biotechnology poses a
significant threat to altering human nature and human dignity, pulling
us toward a post-human future. We do not wish, however, to draw too
sharp a distinction between self-regulation in either sense of the term
noted above, professional and personal, and governing by the state.
Social practices observed and issues outstanding in the implementation of the assisted human reproduction policy indicate the diverse and
varied ways in which bio-policies can be interpreted and actually used
by particular people in specific settings.

The government of Canada has several types of policy instruments
with which it defines and shapes bio-life policy and practice: legislation,

delegated regulations, guidelines, policy statements, research funds, information, and promotion and persuasion. In an earlier chapter, we raised the question of when the provision of information by government becomes the active promotion of a bio-technology product or indeed an industrial sector. How impartial or neutral is the Canadian state then? In that context, what *is* the public interest? In regards to bio-life, a related question is, when does the provision of information, say on genetic counselling, become persuasion or undue influence by health professionals on persons trying to make decisions on abortions and procreation? How autonomous or informed are decisions by women or men or families who often are in circumstances of raw vulnerability? What do personal autonomy, human rights, or informed choices mean in actuality? In raising these issues, our point, building on the work of others, is to signify that information provision and self-regulation are more complicated than is regularly assumed; both are implicated with larger relations of power intersected within myriad social structures and regulatory systems.

How reproductive technologies are used, by whom, when, and why depends on people's material circumstances, family histories, and social identities; on how medical procedures and diagnoses are represented by family physicians, genetic counsellors, and fertility specialists; and on the influence of community groups, the mass media, and social media. In the determination of the self, a great deal more than self-determination is at work. Self-rule of the embodied person interacts with supports and strictures of state organizations and societal arrangements.

7 Power, Changing Biotech Governance, and Extending Democracy

Introduction

In *Three Bio-Realms* we have shown that biotechnology, while no longer a new technology, is still undoubtedly an enabling and transformative one. We have approached an examination of the governance of bio-technology simultaneously as a governance-policy, scientific-medical, commercial-industrial, and cultural-ethical phenomenon. To be sure, biotechnology involves genetic sciences and molecular structures related to food, health, and human life itself; it also unquestionably entails an assemblage of ideas, techniques, actors, products, processes, and resources that construct and are constructed by public and private forms of power and knowledge. We have analysed biotechnology as a political phenomenon, and we conceive the Canadian political community as a structure of power functioning in the multiple arenas of Canadian democracy. The list of biotech products and processes is growing ever larger, and there is recurring dispute about the very nature of what could or should constitute a legitimate product or process in human, political, and commercial terms.

With that wide-ranging orientation, this book had three overall purposes. First, our purpose has been to critically explore the gradual emergence of three shifting, reinforcing, and often colliding governance realms of the bio-economy and bio-society: the bio-food realm, the bio-health realm, and the bio-life realm. Together they now comprise the Canadian biotechnology governance regime, established over the past thirty years. For the regime overall and for each of the three bio-realms, we have mapped and examined the key governance policies, structures, interests, processes, and ideas at play.

Second, we sought to understand the underlying governance dynamics and network-based interests and pressures of biotechnology. In a liberal democracy, these dynamics and pressures seek to support biotechnology and to regulate it in the public interest. They operate equally inside the state and at the broader co-governance levels of business interests, research institutions, NGOs and coalitions, and personal choices and identities by individuals as consumers and citizens and as kith and kin.

We traced the nature of networks as a societal mode of organization linked to an already complex set of interests and as an emerging form of complex accountability and bureaucracy, in part because many of the networks are not just naturally occurring but also policy induced and mandated. We see as well that genes and the human genome are themselves networks in their make-up and functioning. Examples of networks and networked interests in our analysis include BIOTECanada as a reconstituted interest group composed mainly of smaller biotech firms, key features of the work of CBAC and CBSEC, the Canadian Coalition for a Royal Commission on New Reproductive Technologies, the Canadian Coalition for Genetic Fairness, the Stem Cell Network, the Centre of Excellence in Personalized Medicine, and the six Genome Centres across Canada and their GE3LS networks devoted to work on ethical, environmental, legal, and social issues.

With respect to biotech power and democracy, we have shown the development and effects of three kinds of bio-power relationships functioning in and across the three bio-realms. The three forms of bio-power present through the three bio-food, bio-health, and bio-life realms are institutionalized pluralistic power, networked power, and self-disciplined power. These bio-power relations are located within an even more complex set of arenas of liberal democracy, which are, in turn, incessantly subjected to challenge if not change. We have also looked at the partisan and political party influences across the Trudeau, Chrétien, and Martin Liberal governments, and the Mulroney and Harper Conservative governments. This meant the need to probe some of the claims about the adequacy of democracy in the bio-governance regime, when democracy can mean quite different things, including representative Cabinet-parliamentary democracy, federalist democracy, interest group pluralism, civil society democracy, and direct democracy involving citizens as individuals engaged in self-regulation, in focus groups and social networks, as environmentally and health-conscious consumers, and in scientific research and research advocacy. While democracy today is most often identified with elections, political parties, and parliamentary government, we have emphasized the necessity to consider the nature and scope of the Canadian political community as comprising these five institutional arenas, each with specific logics and norms, roles and structures.

Third, the analysis sought to place the Canadian biotechnology governance regime in a context of national and international influences of policies; ideas and processes centred on corporate power and commercial profit, science-based governance, precaution, intellectual property assessments of patentable invention; and risk-benefit regulation in pre-market and post-market phases of production, research, and use. In the bio-food realm, international impacts arose in part from strong U.S. food and agricultural business interests and their close inter-corporate links with Canadian food producers of GM foods. In the bio-health and bio-life realms, U.S. influences were pivotal globally and for Canada. These were seen in the ten-year global genome-mapping project and in fast-forming Internet-based genetic products and research networks that linked to Canadian researchers and health/disease groups.

Economic and scientific ideas and processes we examined centred on research support and funding in the name of an innovation-centred economy and also growing concerns about research ethics and the rapidly growing commodification of knowledge through patents, particularly bio-product and process patents. This involves structures, processes, and ideas that are forged in complex scientific fields, traditional and new academic disciplines and their changing peer-review processes and public agencies, national and international, with particular meanings about how to assess new technologies and products in comparison with well-established ones.

Our analysis has accordingly helped to answer three central questions: What is the nature of biotech governance change? What factors mainly explain such change? And what do such changes tell us about the changing configurations and shape of public and private power as revealed by the Canadian biotech governance regime, as it plays out in a set of complex arenas and kinds of Canadian democracy? Our answers to these questions are captured in the four empirical chapters and are summarized in the six arguments discussed in a final summary way below.

The main empirical contribution of the book is that it provides the first integrated analysis of the political and historical evolution of the three bio-realms over thirty years and hence provides a better understanding of the nature of the overall Canadian biotechnology governance regime and its key changes. The main conceptual contribution is that it develops and employs an analytical framework that systemically explores three elements of biotech governance: the state as supporter and regulator, interests and networked-based democracy, and science-based governance and precautionary governance. The changing interplay among the three elements helps explain the key changes in the overall analytical story and as they are revealed in the main arguments advanced.

In our historical account of biotech policy, the main periods of Liberal and Conservative governments since 1980 show partisan continuity and differences, as biotech policy evolved nationally and internationally. The Trudeau Liberals adopted an initial strategy that overall was pro-food biotech. The Mulroney Conservatives and the later Chrétien Liberals broadened biotech policy to achieve a greater balance of values regarding environmental, health, and stewardship. The Mulroney Conservatives, however, were the driving instigators of Canada's free trade agenda and hence affected biotechnology through its adherence to trade-related norms, including those linked to sound science and stronger patent laws and intellectual property protection.

The later Chrétien and Martin Liberal governments became enthusiastic supporters of genome-related research, mainly via new research-funding bodies and network formation. The Liberals were much slower and sluggish in some of the regulatory policy and governance aspects, including those regarding assisted human reproduction and related bio-health and bio-life aspects. Currently, the Harper Conservative government seems to exhibit both benign support for biotech overall and some concerns about Genome Canada funding. It also abolished the Canadian Biotech Advisory Committee. However, on the regulatory front, the Harper Conservatives' *Blueprint* strategy promises a potentially stronger and more complex regime for post-market, and also life-cycle approaches to the overall regulatory health, bio-health, and bio-food governance system.

This historical picture does not suggest excessively partisan views and differences. It does show that biotech policy emerges in successive governments in direct ways through policies and strategies where biotechnologies are referred to directly in the policy statements, strategies, and discourse. Experience also shows that biotech policy has surfaced indirectly in other policy fields that apply to biotechnologies, along with any number of non-biotechnology industries, products, and processes.

While focused mainly on federal government policy and governance, our analysis makes clear that the more one moves from bio-food to bio-health to bio-life, the more provincial governments and many related authorities such as hospitals, universities, and health professions become crucially involved, as do women's groups and networks, patient and disease NGOs, and individual citizens and carers. This spectrum of provincial and para-public engagement in biotech policy has implications for the relative significance of our federalist form of democracy and its interplay with other arenas of politics practised in the bio-governance regime. It tells us as well that formal structures and centralized authority in state agencies, business corporations, and professional bodies have not disappeared in some post-bureaucratic era; they remain prominent in our social world.

Through a broadly chronological account of the three governance realms, we have shown as well that the actual underlying bio-technologies in bio-food, bio-health, and bio-life have not themselves emerged in orderly chronological ways. Particular biotech products, processes, and research developments were more random, multi-dimensional, and overlapping, and began to surface throughout the three decades being examined, and indeed even before then.

Among the three bio-realms, boundary overlaps also occur. The bio-food realm, clearly focused on food and agricultural products, also has crucial health impacts and related environmental and energy biofuels policy concerns as well. As expected, the bio-health realm is more focused on bio-health and drug products but also is affected in governance terms by reforms aimed at the larger Health Canada food and drug laws and processes, and their recent shift to a greater overall post-market emphasis propelled by national and international ideas and pressures.

The bio-life realm is partly an extension and further elaboration of the bio-health realm but with fundamental differences of its own. Extended sets of players in the system of medical power emerge here, including not only doctors as medical professionals but also related autonomous health professions and the drug industry and university health researchers and changed research granting bodies. These arose in addition because women's groups and coalitions were crucial in putting new reproductive technologies on the national agenda and in influencing the debate about how they (that is, technologies and women's bodies) ought to be governed. The slow, tentative, and uncertain embrace of reproductive technology issues and products became even more complex when extended notions of embryonic stem-cell research, genetic testing, and genome-centred targeted personalized medicine arose with their crucial individual, family, religious, and community senses of identity about life and the human body.

In the context of these overall purposes, we now offer first our conclusions and arguments and related explanations about biotechnology-governance regime change overall, and second, a brief set of final observations and recommendations about biotechnologies and the technical, governance, and democratic challenges, reforms, and other choices they present for Canadians.

Key Arguments and Related Explanations of Biotech Governance Regime Change

Six overriding arguments, and related conclusions and explanations, become apparent across all four empirical chapters in Part Two, including crucially those on the three bio-realms. These arguments emerge empirically, informed by the content of, and interplay among, the three elements in our analytical framework: the state as biotech supporter and biotech regulator, networked biotech governance interests and democracy, and science-based and precautionary governance.

Our first and overall central argument is that biotechnology in its main emerging forms is *altering the boundaries between the public and the private, between the economic and the social, and between the scientifically possible and the ethically preferred.* Since the 1980s, although with longer antecedents, three distinct realms of biotechnology have emerged that are generating a shifting, at times reinforcing, and often overlapping and colliding set of ideas, agencies, interests, and relations of power and knowledge. Biotechnology governance is shaping and being shaped by relations between and among the state, science and technology, the market economy and civil society, as well as by multiple forms of democracy, some classically old and others fairly new, as we struggle to determine the qualities of human life and the meaning of community in the twenty-first century.

Altered boundaries between the public and the private arise from the increasing need for co-governance approaches among the state and businesses, NGOs, universities, charities and other related entities, and individuals, nationally and internationally. These boundary-shifting and blurring features reflect the nature of each new cluster of biotechnology products and processes as well as underlying pressures and changes in democratic governance, including the greatly enhanced power of prime ministers in relation to the Cabinet and the bureaucracy. Public-private boundary shifts emerge regarding clashes about intellectual property, and about how much or which aspects of biotechnology should be patent-based property rights as opposed to being public goods and public knowledge.

Altered boundaries between the economic and social intertwine with the public-private ones as well, and they extend, especially in the bio-health and bio-life domains, to changed and extended notions of what social means. In traditional post–Second World War development and discourse, social policy equated mainly with the social welfare state's health, education, social support, and income-protection policies and practices. In the biotechnology governance regime, however, the social dimension is reshaped and extended, especially as we have seen, by women's groups and networks into diverse meanings regarding the human body, life and death, health, religion, and senses of personal, family, and ethnic identity. This also is why we have emphasized the extended concept of civic regulation in our analysis of the bio-life realm in particular, even though historically some features of it have always existed. In any event, biotech policy issues are certainly important elements in contemporary social policy.

Our second argument is that three forms of power are present within and across the three bio-food, bio-health, and bio-life realms: business-dominated pluralistic power, networked power, and self-disciplined power. We are not the first to argue that biotechnology is a technology of power, but we are more careful in specifying more precisely the different forms of power. These three forms of public power are present in all three bio-realms, though in widely varying degrees. It is not just politics and interest pressures that yield policies or cause policies to emerge. It is crucial to see that the intrinsic nature of policy, in this case biotech policy, helps produce different kinds of politics and different kinds of power and governance arrangements, along with demands for extended and new forms of democracy. As ideal types, these forms of power correspond to state command, sectoral cooperation, and self-conduct.

The business-dominated pluralistic form of power refers to state structures and a narrower pluralistic set of interest group relations dealing with regulation and support. In using the concept of pluralism, we do not imply some kind of perfect balance of representation among interests. In the context of biotech-policy, the system is pluralistic in its formal democratic features and aspirations, but it functions in an extraordinarily institutionalized way so as to favour business interests over others. This emerges most clearly in the bio-food realm, where agricultural and business and research support interests are fairly concentrated and cohesive, particularly in comparison with the mainly dispersed agro-ecology, consumer, local cooperative, environmental, and rural interests (Wittman, Desmarais, and Wiebe 2010). Institutionalized pluralistic power in biotech governance also intersects with the institutionalization of gender relations in federal laws and regulations, social beliefs and family practices, and the practices of official politics and grassroots politics. This form of power exists to some extent in the bio-health realm, where drug and biotech companies had similar cohesion but faced a much more diverse set of networked interests in the form of the medical profession, researchers, patients, and disease groups in a fast-changing system of medical power.

The networked form of political power refers to state-society/economy relations and co-regulation and self-promotion by private and research interests. The bio-health realm most exhibits this mode of power, in part because, although drug and smaller biotech companies are a central interest, there are literally dozens of networks of science, scientists, and medical practitioners involved, also now encouraged and mandated by policy to be increasingly network-like and partnership-based in the way they work and are funded. As well, networks are technically inherent in the actual nature of the scientific-genome structure of genes and links among genes and with diverse complex environments.

The self-disciplined form of power refers to governance of the self by the self, influenced to be sure by actions of the state and other actors, yet in significant ways governed by highly personalized forms of power beyond what the state can actually control or have power over. Crucially, this includes gender power, where women in particular seek to empower themselves vis-à-vis men with regard to numerous aspects of reproduction and the caring of families and friends. We see this particularly in the analysis of the bio-life realm, and it is also a feature of the bio-health realm. Overall the self-disciplined form of power relates to civic regulation but also self-regulation defined partly to be regulation of the self by the self. This involves therefore the private behaviour of consumers, citizens, and, more generally, individuals, particularly 'women' as a category, which we deconstruct as persons in certain social roles as 'embodied actors,' such as infertile partner, expectant mother, aging parent or grandparent, or person with a developmental disability or other impairment, among others. An important element here is 'embodied justice': the legal construction and regulation of embodiment, with biomedical, disability, ethical, feminist, and human rights perspectives on new predictive genetic testing and notions of the 'healthy embryo' (Nisker et al. 2010). Institutions in which this third form of power operate include families and related kinship relationships, women's organizations and support groups, health clinics, physicians' offices, and hospitals, with the body as an object of power and related knowledge and research (McLaren 2002).

The third argument we make is that fundamental political realities explain why the Canadian biotechnology governance regime does not have a central point of governance within the state and why it has rarely been a high political priority. This seems paradoxical, given the evident transformational nature of biotechnology.

The first part of the paradox is that biotech governance has become more complex and even impenetrable over its recent history and that it still has no obvious single institutional centre. From the outset, bio-governance has been constructed in its various realms as a multi-agency appendage and set of subunits to federal departments and agencies with earlier, much larger, non-biotechnology mandates. In part, this was initially because biotech industry interests did not want the new technology to be seen and cast as new or threatening but rather simply as novel adaptations of existing technology, cast as novel foods. Later manifestations of agency complexity built on this initial design, then grew markedly as new products and processes appeared.

The second part of the paradox is that biotechnology has not reached the pinnacle of the federal agenda, since it has scarcely been mentioned or given much of a profile in Throne Speeches or Budget Speeches since 1980. These are the central agenda-setting occasions for prime ministers and ministers of finance. We noted that of the thirteen total mentions in the fifty-four total speech documents across thirty years, ten were during the Chrétien-Martin years, but even here they were mainly subsumed under their broader innovation, S&T, and health-policy agenda-setting and thematic narratives. This thirty-year story also shows the types of language or discourse used. One type is generic or high level, such as calling biotech 'strategic' or 'enabling' and only one of several such technologies. The second and somewhat more common type of discourse in these agenda speeches is specific to one or other realm or industrial sector, such as life science, medical research, agriculture, bio-fuels, or pharmaceuticals.

Biotechnology functions in Canada in the subdued middle worlds of political life and governance for a number of linked reasons. These have been contrasted in part with a higher-profile political presence at times in Europe and the United States. In Europe, bio-food opposition was strong and sustained, partly because of the relative absence of bio-food producers. There has also been a higher periodic presence in the United States, in part because of the stronger role of religion in U.S. politics in general and in bio-life issues. The posited reasons and features also include the argument that ministers and politicians may not quite know how to deal with biotechnology in a raw political sense; the bio-world involves complex kinds of science and technology that are beyond the capacity of non-scientist politicians to discuss and communicate comfortably and with clarity; and, because bio-technologies and their governance deal with boundary-spanning notions about the public and private and the social and economic, they yield further discomfort for politicians who have to try to explain complex forces and values in an otherwise sound-bite age of political discourse.

In a raw political sense, there were few political positives for them in the initial bio-food realm. The discourse of Frankenfoods and unnatural food was best avoided, hence most ministers just stayed away and left issues and subdued support actions to ministers of agriculture or to MPs from mainly Western Canadian prairie agricultural/food regions.

In the bio-health realm, more political positives operate regarding potential health benefits. Nevertheless, there were still political challenges and some real political negatives, given the sheer speed and complexity of change in product volumes and in claims about a bio-health nirvana. It was also even more difficult for politicians to understand and then discuss the complex science and technology aspects, let alone venture into details about the nature of innovation or the obscure technical world of intellectual property, patents, and what constitutes a legitimate invention.

In the assisted human reproduction and genome-centred aspects of the bio-life realm, core political concerns about how to deal with this even-broader realm in everyday social life, discourse, and governmental sound-bite responses became even more difficult. Issues about the nature of life, mapping the genome, genetic testing, personalized medicine, reproductive technologies, and extending and redesigning life challenged the senses of identity of women versus men, and of lesbians or gays versus straights. Such issues also challenged the identity of individual politicians across political party spectrums whose normal fault lines regarding social versus economic policy and the public versus the private did not enable them to know quite what to do or say about biotechnology, even when softened by supposedly comforting umbrella notions of a dynamic bio-economy and a hopefully beneficial bio-society.

It is also important to comment on a question of power that emerges from the discussion above that biotech policy and biotech governance has not had high profiles in the national agenda. Which interests benefit from its continuous presence in the nether or middle regions of political life, and what kind of democracy does this reveal or constrain? One interpretation is that this state of affairs benefits bio-business interests and certain health-research interests as well, in that policy and governance does not need to deal with full-scale national debate, as one sees more frequently, albeit never perfectly, in policy realms such as economic policy, social welfare policy, environment and energy policy, and law and order policy. It also implies that representative parliamentary democracy is not seen as having anything other than a marginal role.

Three Bio-Realms shows that Parliament's role does seem very indirect but it also demonstrates that Parliament is consequential for biotech policy and governance.[1] Rather than a robust policymaking body on biotechnology, the Canadian Parliament has been a policy-influencing body within the broader context of growing executive dominance by prime ministers and Cabinets. No mere or automatic rubber-stamp of government policy on biotechnologies, members of Parliament and senators have regularly raised questions, presented concerns, conveyed the views and preferences of their constituents and other Canadians, and offered alternative courses of action, such as on genetically modified foods and the labelling of products. Specific roles and the actual effects of Parliament have varied over the 1980 to 2010 period, which has been our prime focus.

Parliamentarians played a noteworthy role in reviewing a draft of legislation (Bill C-47) on genetic technologies through 1996 and 1997; consulting with an array of organized interests from business, science, women, and faith communities; providing a space for witnesses and submissions; and making recommendations for amendments, some of which reflected in the subsequent legislation. In reality, anticipating the opposition or collaboration of political parties in Parliament and, by extension, the dissent or consent of organized interests in society is a central feature of policymaking for the federal government. This form of policy development by anticipation is an invaluable technique for managing contentious issues and designing rules on topics such as surrogacy or pre-conception arrangements. In addition, the Assisted Human Reproduction Act contains a provision for parliamentary review of the act by a committee of the House, Senate, or both, and for tabling any report from such a review for further debate and possible action. With the formation in recent times of the AHR Agency, the CFIA, and Genome Canada, Parliament has an opportunity as well as constitutional obligation, we believe, to scrutinize the conduct of these and other entities in the Canadian bio-governance regime.

A further related concluding question is what the recent history of biotechnology – understood as scientific developments, economic applications, and social issues – reveals about the ability of Canadian government agencies to formulate policies and implement regulations that govern bio-food, bio-health, and bio-life. We offer three conclusions on this question.

First, the Canadian state indicates capacity in biotechnology by developing and refining policy frameworks on science and technology and on regulatory governance; by organizing and managing consultations with a range of interests and democratic forms on biotech policy; through establishing core organizations with mandates directly bearing on biotechnology, such as CBAC, CFIA, Health Canada, and Genome Canada; by having Cabinet and public service take certain policy actions, even while legislation has yet to be approved; and through participating in the global arena, in promoting and signing international agreements and protocols on biotechnology and science-based governance.

Second, biotechnology politics and policymaking in Canada bring to light important factors that constrain or pressure the ability of the federal state to regulate in this governance regime. Consider in the bio-food realm that in the debate over mandatory versus voluntary labelling of genetically modified food products, industry interests have prevailed over consumer interests to limit the use of standards and laws to prescribe labelling.

Consider in the bio-health realm that medical and scientific communities exercise notable influence, not least through invoking the principle of professional autonomy and also peer review and thus the necessity for the self-regulation of practice.

Consider in the bio-life realm that the federal government's core leg-islative initiative on assisted human reproduction faced a major and partly successful constitutional challenge from the Quebec government over the question of division of powers. In these cases, we see the direct influence of specific elements of capitalism, professionalism, and feder-alism upon the Canadian state's capacity to govern biotechnology.

Such instances of economic, social, and political constraints on a government's rule-making ability are not unique to biotechnology, of course. Indeed, we expect the embedded state in a liberal democracy and capitalist economy to operate within imperfect networks of orga-nized pressures, mixed expectations, provisional choices, and uncer-tain opportunities and constraints. The discussion in previous chapters does show that biotechnology is a multidimensional phenomenon; and so, too, is state capacity to regulate and govern biotechnology.

Third, the fact that there is no obvious institutional centre to the fed-eral governance regime and that biotechnology has not been a high priority on the national agenda does not necessarily signify a lack of capacity by the Canadian state. We contend that these characteristics of biotechnology policy and governance say much about the complexity and contested nature of our political economy, of science and public morality, and of the self. The Canadian state itself, we have argued, is conflicted by its institutionalized commitment to both support and regulate biotechnologies, often by different segments of the public sec-tor with different levels of authorization and resources.

Our fourth, fifth, and sixth arguments build on the first three already summarized above. In a complementary manner they allow us to see in more particular ways the roles and interactions among the three ele-ments of our analytical framework and how together they help explain governance differences in the three bio-realms.

The fourth argument that emerges from the analysis is that the state's need to support and to regulate transformative technologies such as biotechnology has resulted in ever-broader notions of what support activity and regulatory activity involves and also the discourse regard-ing how to speak publicly about these partly conjoined, partly separate, and often conflicting actions. There is no industry in Canada that is not simultaneously supported and promoted by some agency of the state and also regulated by some part of the state. Thus biotechnology is not alone in having these core features of supportive governance and regulative governance forged, debated, and changed. These tasks and challenges immediately raise concerns about the need for regulatory independence and even institutional firewalls within the state; about positive support for industries, firms, and researchers generating new products and processes; and about democratic horizontal coordination and joined-up government within elected representative Cabinet-parliamentary government and democracy.

Our argument on the support and regulatory tensions in bio-governance draws on chapter 3, where the focus was on federal research and related support agencies and foundations, with reference also to the regulatory roles played by these granting bodies and policy advisory and coordination agencies. In the three bio-realm chapters, the analyses also show how support actions and supportive discourses develop and change within and around regulatory mandates as well.

Across the last thirty years, there have been ever-broader notions of what constitutes support and regulation. This broadening has come about in the use and deployment of policy instruments such as spending, taxation, regulation, and persuasion, in agency mandates and in basic political discourse. These changes reflect direct biotech policy influences and changes propelled by other policy fields that indirectly but still significantly affect the bio-realms. Aspects of supportive governance were reflected in increased research spending via several research-granting agencies, only one of which, Genome Canada, was clearly focused upon biotech. Support was also manifest unambiguously in the early 1980s biotech strategy but was then qualified significantly in later Industry Canada and CBAC statements, strategies, and studies. Articulation of an overall innovation strategy in the late 1980s and 1990s also provided somewhat more positive and legitimate political cover via a genuinely important economic policy challenge. In bio-food for example, the industry did not have to speak of biotech and justify it only on agri-food or novel food grounds, but rather could lobby on the broader grounds of innovation as well.

The bio-food, bio-health, and bio-health realms and their respective agencies, products, product volumes, and rule-making norms, laws, rules, and soft-law were partly layered next to and on top of each other. Analysis of the bio-food realm shows that bio-food arrived into a Health Canada food and drug regulatory system where post-thalidomide 'safety' mandates and values focused on pre-market assessment were dominant. The then newly created CFIA was similarly motivated and mandated as an inspection and enforcement agency made arm's length from Agriculture and Agri-Food Canada. Our account of the bio-health realm draws out the ways in which both bio-health and overall drug regulation was changing to encompass a needed new and extended focus on post-market review of products and processes and also, given greatly increased product volumes, on a need for regulatory efficacy and speed in comparison with other regulators in competing countries.

Within the regulatory broadening dynamic, there also came a far greater explicit discourse on the fact that health-regulation mandates were (and always had been) centrally involved with risk-benefit regulation. The benefit of mandate discourse, not to mention the underlying belief (and frequent hyping) of bio-health benefits, was easily seen as a supportive set of actions and changes as well, and hence, in the eyes of critics, a weakening of the purer safety mandates of regulatory logic, discourse, and practice.

By the time that the bio-life realm emerges in recognizable institutional form, the support and regulatory governance forms and nexus take on still further complicated characteristics. Chapter 6 shows, through its initial focus on Assisted Human Reproduction Canada and its parent Assisted Human Reproduction Act, a set of paradoxical features. On the one hand, the decade-long delay while first a royal commission deliberated on this aspect of bio-life and then slow law-making itself from 2004 to 2007 meant that deferral itself was partly a form of support. New bio-life practices and processes emerged in the health and social system, including new genetic testing products and processes, evolving self-regulation, and civic regulation modes of governance.[2]

Strong sets of values, many of them conflicting, extend the meanings of governance support and regulation. The deeply felt views of women and the often contrary views and religious convictions held by many Canadians on family, gender, disability, human dignity, and the morality of inclusion make this a politically intricate and socially charged regulatory domain. In addition, policy debates are infused with strongly entrenched positions and often diverse legal principles, self-understandings, and economic interests. This means that there is an arduous challenge in exactly what to say in public debate and in drafting legislation or regulations in this realm.

This discussion leads to a question of how biotechnologies are directly portrayed in political talk and public discussion in biotechnology and in other policy fields that apply to biotechnologies. In regards to other policy fields, we mention trade, research ethics, consumer protection, innovation and technology, intellectual property, and health care. While not wishing to emphasize simple and rigid binaries in political language and interests, there does appear to be a pair of internally coherent and mutually competitive discourses from these other policy fields, as they pertain to biotechnology, that look something like this:

- The dominant discourses from policy fields such as trade, innovation and technology, intellectual property, and regional economic development tend to favour notions of promotion and support, commercialization of research and development, and property rights over science discoveries. Principles and practices of science-based governance and international trade norms reinforce these notions, especially notions of science-based governance that privilege R&D, invention and modernization, and related scientific activities. We have also noted in various parts of the book that medical and scientific communities tend to advance a discourse of autonomy, peer review, and professional self-regulation for their activities in relation to the state as well as express an optimistic view of the general benefits of biotechnologies. Overall, this is a particular expression of civil society democracy in the bio-governance regime. In other arenas of democracy, the dominant discourse emphasis is on Cabinet-parliamentary government and federalism. To the extent that this perspective endorses direct democracy, it tends toward an entrepreneurial analogy, viewing citizens as consumers in a market society looking for and wanting new and improved goods and services. Here a variation of 'rights talk' champions the right of individuals and families to the latest technologies, products, or processes that science has yielded. In sum, this is supportive and business-oriented, science-based governance.
- In comparison, the principal discourses from such policy fields as research ethics, gender policy, consumer and environmental protection, genetic counselling, and health care tend to favour notions of regulation, an ethic of care (over commodification), and view science-based governance in terms of scientific applications as public goods. Important principles here include informed consent, free choice, quality of life, and personal autonomy. An array of interests and civil society groups are engaged in a safety and risk discourse that emphasizes the uncertainties and hazards associated with many biotechnology processes and products. This discourse is often expressed through principles of precautionary governance and environmental policy norms. Other groups are also involved in an

ethical discourse that underscores morality and human dignity and norms of human rights. In total, the democratic arenas highlighted here are interest group and civil society democracy along with more conventional forms of Cabinet-parliamentary and federalist democracy. Direct democracy in this discourse links with identity politics: the mobilization for social change and policy reform by individuals and groups disadvantaged as the result of intersections of age, disability, ethnicity/race, gender/sex, or infertility, among other characteristics. Overall, groups and interests here are pre-disposed to a deepening of democratic processes and practices. In sum, this is regulative and societally oriented, precautionary-based governance.

We think that the former discourse has, on the whole, been the more powerful and influential one and that, as a consequence, there is a need, as we show in the final section of this chapter, for a rebalancing of discourse and power in a way that gives far more salience to the latter kinds of discourse, democracy, governance, and policy change.

We can see even from a summary overview how biotech policy and biotech governance inescapably concern converging and often oppo-sitional sets of ideas and hierarchical interests about our economic, social, and personal worlds; about our images of humanity and person-hood; and about how we should live together. This is one of the basic reasons and surest signs that biotechnology is political. It entails indi-vidual actors and institutional interests expressing certain values, tak-ing sides on given issues, making calculations of relative benefits and risks, and supporting and opposing particular ideas and policy propos-als. Biotechnologies have altered our notions of food and nature, health and illness, and human life. In this sense, we experience a 'geneticiza-tion' of politics and public discourse along with a politicization of the life sciences.

The fifth argument advanced is that the increased networked nature of the biotech governance regime's public-private interests and indi-vidual citizen identities makes it increasingly difficult to judge claims about whether the regime overall is democratic, and if so, what kinds of democratic principles, and what kinds of outcomes and accountabil-ity are required. Networked biotechnology governance, as a descrip-tor of actual working practices and relations of power and knowledge, can make some claims to greater and higher democracy and inclu-siveness, but it can also produce impenetrable transaction costs and accountability mazes such that it can produce a kind of complex, obtuse democracy in stalemate.

This argument emerges conceptually from our analytical framework that deals with networked bio-governance interests and democracy and empirically from the overall observed story. Key interests involved across the three bio-realms include business, industries, and financial capital; NGOs, women's organizations, and other gender and civil networks; science, medicine, and technology; and departments and an assortment of arm's-length agencies characterized by partly separate interests from government as a whole.

The notion of these interests being increasingly networked is true in general, especially because network-style governance has been explicitly advocated across the last three decades and have been contrasted with other modes of social and economic organization such as markets and hierarchies. But the analysis overall shows, as does the conceptual literature, that networks can be both naturally occurring, as in the normal conduct of research, and peer review among scientists, including among bio-researchers. Networks are also more policy induced or required. Research funding and new bio-related agencies such as Genome Canada are network based as a matter of policy, including new levered forms of funding that required partners with money and expertise to be found. This is also the case with the CIHR, the CFI, and of course the Networks of Centres of Excellence and their biotech funding.

Across the three bio-policy realms, the networked nature of interests varied but also grew more complex. In the bio-food realm, networks as linkages of political interest, discussion, and conflict were more confined to industry groups and firms and financial capital on the one hand, and consumer interests on the other. They, in turn, were networked with a small set of agencies centred mainly in Health Canada and the CFIA. Environmental interests and networks, including Environment Canada, were active but were also left mainly to the margins of governance, certainly in the basic design of the core bio-food regulatory realm.

In the emergence of a full-scale bio-health realm, networks were much more explicit, encouraged, and required. Patient, disease, medical, and research interests were more network based in their varied internal structures and in their need for alliances of support and expertise. The expansive shift from pre-market to a greater post-market regulatory focus could be achieved only through complex networks of players and interests in the larger medical-health system and in the context of the higher volumes of bio-health products.

In the bio-life realm, networked bio-governance interests were even more endemic and indeed needed. Networks of women's groups, as well as other non-governmental associations and interest groups serve a number of roles: knowledge production and evaluation, the promotion of some specialized products and therapies, and rule-making and implementation. Medical and other health professions constitute an important networked arena of non-state actors for bio-life regulation, information, advice giving, and interventions. The result of this dense community of interests and institutions is a fluid, multi-level context of regulatory governance, with the concurrent realities of practical complexities, diverse and often divergent messages, inherent tensions, and abiding anxieties over the future of human nature.

Our fifth argument includes a wider observation about how increasingly networked interests adversely affect our ability to assess claims about democracy in the governance of biotechnology. We have already referred to bio-governance exhibiting three kinds of political power within and across the three bio-realms. While these systems of power are important to observe, they are not fully equivalent to the different and often conflicting normative criteria of democracy.

Biotechnology governance in Canada has played out in a larger set of arenas and related values about democracy. We discussed these in relation to five such arenas: representative democracy (in Canada's case, elected Cabinet-parliamentary government); federalized democracy; interest group pluralism; civil society democracy; and direct democracy involving individual citizens engaged in self-regulation, in focus groups and social networks, and as environmentally and health-conscious consumers.

Bio-governance includes all of these to some extent across all three bio-realms, yet they emerge with greater salience in some more than others. Representative government, in our case with a national government focus, is clearly present in each of the three bio-realms in law-making and regulation-making, not to mention overall foreign and trade policy. At the same time our analysis has shown that biotechnology and bio-governance has only periodically reached the top of the normal agendas led by the prime minister and finance minister, with the implication being that there is the paradox referred to in our third argument and that there is some kind of democratic deficit in elected representative democracy. Parliamentary committees have had important impacts on biotechnology policy and debates, even though not well known or reported on in everyday political coverage.

In the Canadian context, as in other federations, a form of federalized democracy is present without doubt. Our analysis has focused overall on federal bio-governance, but the more one moves across the three bio-realms, the more that federalist democracy arises and hence the roles, views, and jurisdiction of provincial governments ultimately cannot be downplayed. The bio-health regime exhibited this arena of democracy quite starkly, especially regarding the growing need for new and high-volume bio-health products to be paid for by the provinces under provincial health-care plans. The bio-life realm also showed federalist concerns when Quebec successfully challenged through the courts, by way of a reference case, several key sections of the federal Assisted Human Reproduction Act.

Interest group pluralism as a democratic arena and norm emerges most clearly in the bio-food realm. In the bio-health and bio-life realms there is a much more complex interplay among such pluralist groups and among the two remaining broader democratic arenas of civil society democracy and direct citizen-focused democracy. In the bio-food realm and in our discussion of CBAC, it was apparent that some groups that saw themselves in broader civil society terms felt themselves to be outside the pluralist arena, which they characterized as being too business-dominated. These included many environmental groups as well as rural social movements.

As for arenas of direct democracy, all three bio-realms exhibit direct individual citizen involvement. In bio-food, these took the form of citizens expressing their concerns as consumers and about consumer information and choice regarding bio-foods. They were also involved in focus group approaches to consultation that went beyond group affiliations. In bio-health, citizens functioning as patients and in smaller disease groups loom large. And in bio-life, citizens seek direct forms of democracy regarding their views about life, death, the health of their own bodies, and varying kinds of identity, both their own and those of their families and communities.

The fivefold nature of these arenas of democracy itself makes assessments of bio-democracy difficult, often because advocates of one arena are frequently silent about the claims and legitimacy of the others. Related areas of democratic accountability and participation become increasingly complex and opaque precisely because biotechnologies are so varied and numerous at the product level. Furthermore, biotechnologies involve, as we have stressed, both direct and indirect policy impacts across diverse overlapping policy fields, from biotechnology as such, to health, trade, agriculture, innovation, science and technology, and intellectual property.

Trust in biotechnology democracy is a many-splintered thing. Higher trust in democracy, a concept discussed early in our analysis, is the seemingly elusive outcome of a constant interplay among the five notions of democracy. This interplay makes trust a fluid and contested resultant of a number of arenas, interests, norms, and processes. Canadians' trust of Cabinet and parliamentary democracy is more problematic, given declining voter turnout, scepticism towards politicians generally, charges of centralized control in the hands of the prime minister, and the recent period of minority Parliaments. Federalist democracy, by definition, embodies divided loyalties across divided jurisdictions and political communities, inevitably adding pressures of fragmentation and tension on any project of building trust within the body politic. Turning to interest group pluralist democracy, although perhaps public trust in science and medicine is not as high or readily accepting as a generation or more ago, it remains solid and widespread, especially compared to other institutions in Canadian society and economy. Regarding civil society and direct forms of democracy as they relate to biotechnology, the emergence of new groups and the enhanced mobilization of existing groups shows a level of trust or faith in representative democratic processes to a degree, while at times perhaps suggesting also the loss of faith in other organizations in the political system to articulate and advance cherished values and beliefs by particular groups and communities.

Across the three bio-realms, we suggest that the form and level of trust varies, with trust perhaps highest in health products and lowest in certain food products, and bio-life products somewhere between. The contingency of trust is apparent by the fact that the confidence and support Canadians have for bio-techniques varies within each realm too. In this sense, trust attaches to the specificities of local needs and personal applications.

Our sixth and final argument is that science-based governance norms and principles still underpin the biotech governance regime more than do the norms of precautionary governance. These principles are often seen as polar opposites and are not universally used, but these two different approaches exist and are also converging and colliding in key ways.

The notion of science-based governance has itself taken on extended meanings and needs, including evidence-based and intellectual property invention assessment. It applies differently to law and regulation-making on the one hand and product assessment on the other. Moreover, in product assessment, it applies not only to multiple pre-market regulatory-product approval cycles but also full-product life cycles, to the overt use of research funding to support the technology, and also to increasingly complex post-market phases of biotechnology use and hence the need for other expanded knowledge-based forms of review.

Precautionary governance norms certainly are an important part of the analytical story. The precautionary principle emerges from environmental ideas, laws, and policies, international and Canadian. However, precaution as an idea and a human instinct may be induced and produced not just by the precautionary principle as such but even more so by the sheer complexity of the biotechnology governance regime. A central feature of this complexity arises from the folding in of other key values centred on ethics in biotechnology research and in the use of biotechnology in particular products.

In our discussion of the third element of our analytical framework, science-based and precautionary governance, several features were noted as falling within its domain: R&D and RSA as different kinds of research and regulatory monitoring science; evidence-based analysis, including cost-benefit analysis; intellectual property application assessments for patents; peer-reviewed science and processes among and within networks of science, including the social sciences; and the precautionary principle. We linked science-based and precautionary governance to debates about and searches for proper arenas of formal technology assessment. The notion of precaution, as we have discussed, has two dimensions: first, precaution as possibly emerging through adherence to key stages of science-based risk-assessment decision-making, regular engagement with diverse interests, and networks; and second, precaution as a kind of ill-defined resultant from all of the decision and assessment processes in place, including some that might be considered partly, albeit very imperfectly, as technology assessment arenas.

Evidence from across and within the three bio-realms confirms the cumulative presence of these features and of their greater presence in some bio-realms more than others. The analysis of the bio-food realm shows that the heart of science-based regulation resides in the assessment of food products where the core interactions are between Health Canada front-line science assessors, and scientists from the applying firms. These assessments and interactions were criticized by bodies such as the Royal Society of Canada, which deplored the absence of a proper precautionary principle, the use by Health Canada (and other countries) of the concept of substantial equivalence to existing novel foods, and the absence of proper and transparent peer review. Similar concerns were raised in different ways regarding the CFIA's environmental assessment of plants with novel traits.

By the time the bio-health regime emerged, science-based governance was still central, but this time, the conduct of the core front-line science assessment was augmented by the efficiency needs and time demands propelled by pressures for innovation-driven 'smart regulation,' especially given both new health products and high volumes of them. Moreover, because of the greater importance of patents, product assessment became multiple cycles of evaluation.

Patent assessments, as chapter 5 has shown, were formally more technically focusing on the three criteria of novelty, efficacy, and non-obviousness. These assessment processes and related decisions were themselves more contentious as bio-health products were raising more and more concerns about what it really meant to 'invent' when genome-related 'products' and inventions were involved. Crucially, these concerns raised issues regarding what the limits of property rights and hence the commodification of health and life ought to be. The bio-health regime's extended ambitions regarding the post-market also meant that science-based governance increasingly involved related science activities or RSA rather than just R&D. In addition, it involved the kinds of broader and complex networked medical expertise and disease groups and patient self-diagnosis via Internet information.

In the bio-life realm, these extended ranges of science-based and diverse knowledge-based governance and also precaution are even more evident. Partly this is because the bio-life realm includes individual citizens participating by means of seeking knowledge, information, and research on reproductive products or genetics; participating in and interacting with genetic communities; making choices about personalized medicine; deciding to make or to obtain an egg or semen donation; intermingling church and state in personal discourse and action about reproduction policy; and in making decisions on who will be born, determining identity-based life choices with implications for human diversity and genetics. Individuals, then, are not solitary and unitary subjects, but actors in the flesh with potentially numerous roles and discourses in different circumstances over the life course.

Biotechnologies thus increasingly interconnect to embodied subjects, for men and women in numerous social locations. In at least two senses, human bodies and biotechnologies overlap across all three of the bio-realms we have examined. In one sense, the body is a material location for biotechnology research, products, and services dealing with food, health, and human life. In a second sense, bodies and biotechnologies have common ground across the three bio-realms, in that bodies bear the imprint or effect of cultural beliefs on individual and collective understandings of high-quality food and good eating, personal health and well-being, and of families, healthy babies, and parenthood.

To a certain extent, the ways in which biotechnologies relate to human bodies differentiate by type of bio-realm. In the case of the realm of bio-life, to take one example, the embodied subject is a source of biotechnology products and related activities for the reproduction of human life itself, in a way that is distinct from the food and health realms. What is more, in the bio-life realm the human body is a means by which individuals are constituted, and identities are produced or not within families.

While every realm of biotechnology entails connections between science-based governance and politics, in the interplay of human bodies and biotechnologies, each realm has a distinctive pattern of institutional structures and relations as well as social and ethical implications. Within and across the Canadian state, society, and economy, regulatory governance for biotechnology is multilevel and multi-sectoral and involves multiple perspectives. This means that it is extremely difficult to envisage and design democratically or practically any single type of idealized arena of technology assessment. Both the systems of bio-power and the multiple arenas and diverse criteria of democratic politics make it impossible for one-size-fits-all solutions. This calls for multiple processes and the open contention of ideas, which, in our view is welcome, since 'democracy is something uncertain and improbable and must never be taken for granted. It is an always fragile conquest that needs to be defended as well as deepened' (Mouffe 1993, 6).

Biotechnology Governance Challenges and Extending Democracy

Several biotechnology governance challenges arise from the varied technologies themselves and from the nature of biotech governance.

We draw concluding attention to five such challenges and the policy and governance reforms they require as they emanate from the bio-food, bio-health, and bio-life realms.

First, with regard to bio-food, our governance focus was on bio-foods consumed directly by human beings as consumers. We dealt briefly with aspects of plants with novel traits where the concern was with the potential unconfined environmental release of such bio-plants. In the future use of bio-food, there are very real concerns about how much consumers and users really know about the safety of such products, especially regarding environmental impacts and about consumer information. On the latter point, we believe there is a strong continuing case for the compulsory labelling of bio-food products so that consumers can choose or not choose to consume GM products.

On the issue of environmental impacts, these were among the initial concerns of the Royal Society of Canada in its assessment we discussed in chapter 4. These concerns undoubtedly remain important, all the more so because there are increasing claims by bio-food producers that such impacts need not be a concern, and that indeed bio-foods and plants can be a big part of global success in producing sustainable agriculture. All of this suggests that the environmental impacts of biotechnologies, in particular regarding plants and their food linkages, need to be given a sounder and more transparent research-related governance base, especially regarding assessments of longer-term and still largely unknown effects on Canadian farms and farmers, and also consumers. Not all of the Royal Society's arguments about the need for peer review on individual products assessment are necessarily valid or even practical, but there is certainly a case to be made for the Royal Society of Canada or related bodies to conduct periodic public arm's-length assessments of such impacts and reforms that may be needed in the light of such impacts, positive and negative.

Second, it is important to issue a cautionary note about the underlying blueprint for reform agenda in Health Canada and the federal government's policy and regulatory agenda-setting gaps overall. The Harper-era blueprint agenda builds in part on directionally similar changes planned by the Martin Liberals and also by changes to the U.S. law that governs the role of the FDA. We have given this Health Canada reform agenda focused attention, since the blueprint reform seeks to balance the pre-market phases of product regulation with a new and extended focus on the post-market and indeed on a full life-cycle system of product regulation.

This reform is a valid one, but it is still more an aspiration than a reality. Its potential success depends on many factors, including a promised but not yet delivered firm statutory base, considerable human and scientific data resource commitments in an period of fiscal austerity, and a fostering of extremely complex networks of reporting and monitoring that will be much harder to design and defend as a workable and transparent system that Canadians can trust and rely on.

We have also drawn attention to agenda-setting gaps in other senses. Economic, ethical, scientific, and social issues escape full treatment and debate simply because biotechnology, despite its path-breaking importance, has very infrequently reached the highest level of federal agenda-setting as determined by prime ministers and ministers of finance. Many of the linked regulatory issues and gaps call out for a much more explicit and formal national regulatory agenda to deal with all of the related biotech dimensions so that they receive concerted attention along with other regulatory matters. These include the health regulatory blueprint issues, patenting rules, medicare, consumer regulation, the regulation of ethics, and environmental rules of many kinds.

Third, from the bio-health realm of products and processes we have seen arise high volumes of new biologics and genetic testing products aimed at small market niche sets of patients and diseases, extending ultimately to new forms of personalized medicine and genomics. Consequently, a key issue for Canadians, examined partially in chapter 5, is this: At the margin, how many of these should actually be paid for under medicare, under national or provincial health insurance? And what kinds of agencies and forms of democracy and expertise are needed to grapple with these tasks? Many of these bio-health products have high financial costs and also potentially great political controversy as to whether they are truly life saving, life extending, or life enhancing, where matters of life and death emerge starkly in diverse forms of media and Internet scrutiny and self-diagnosis, on a potential product-by-product basis. New biologics and genetic tests also raise the need for more open bio-health and wider health-care forums to debate issues about particular products, their funding or non-funding under medicare, and whether Canadians will tolerate what may turn out to be a particularly controversial form of heath-care lottery that produces different regional and intra-regional responses and outcomes and thus generates different senses of health-care equality and equity.

A fourth and related current governance issue spanning bio-health and bio-life concerns the high financial expenses, aside from emotional costs, families incur for infertility treatments, surgeries, and fertility drugs that are not covered under the public health insurance plans in most provinces. Families have tried for IVF funding through actions in the courts and human rights tribunals, by petitions and demonstrations, and lobbying through newly formed coalitions. Ontario has provided some funding for IVF treatments since the mid-1990s and been under concerted pressure to extend coverage. Quebec introduced in 2008 a refundable tax credit for expenses related to artificial insemination or in vitro fertilization, and in 2010 Manitoba similarly introduced a fertility-treatment refundable tax credit. The result is a classic pattern in Canadian federalism of patchwork public funding for, and accessible provision of, assisted reproduction services across the country.

Concerning the greatly extended realm of bio-life products and processes, the fundamental challenges are almost without bounds and are undoubtedly not yet internalized or appreciated by most Canadians. The product and medical-practice realms and choices within both bio-life and bio-health enter a Canada with an aging population but also a relatively rich one. What is more, bio-life products and practices involve diverse viewpoints and values by a Canadian population whose ethnic and religious mix is also changing rapidly. If bio-life product and technology choices involve intensely diverse beliefs about life and humanity, and also diverse kinds of personal identity, then they are bound to be both personally intense and politically complex.

Canada has had some arenas for debate about these bio-ethics issues and choices. These have included the Genome Canada and regional genome centre GE3LS processes and discussions, and of course they were a part of the 1989 to 1994 process of the Royal Commission on Reproductive Technologies. However, our analysis suggests that these deliberative processes, while valuable, are episodic and seem quite marginalized. Some larger permanent arenas linked much more closely to both parliamentary government and federalist democracy are needed, since these bio-ethics issues will become ever more present and complex in the governance of biotechnology and of Canadian society.

Our final point is that biotechnology has changed and continues to change the nature of science and invention, and how one assigns rights to intellectual property on the one hand and public goods science on the other. In contemporary societies, biotechnology is implicated in notions of democracy and social identity as well as scientific inquiry and commercial applications. A much more open public policy debate is needed in Canada about intellectual property policy and biotechnology, particularly regarding patents. For far too long, patent and larger intellectual property governance and policy has been debated and determined in the technical back rooms of government, business, and the legal and IP profession, and been dominated excessively by a protection and property rights ethos.

As a matter of reformed policy and governance, the evidence increasingly indicates that patenting rules and regimes must shift away from their two-decade-long focus on property rights and commodification to a new focus where larger proportions of invention and knowledge are left in the public and social domains. Biotechnology is changing the nature of what patents are and ought to be. That is why there needs to be a more open arena of debate and governance in Canada, because of biotechnology patent issues, and because analyses of innovation are properly much more sceptical of patenting as a single-variable cause of innovation, when clearly it is not. Publicly available knowledge has equal claims as a significant driver of economic and social innovation.

In view of these bio-governance challenges, biotechnology will further change the boundaries of the social and the economic and of public and private power in Canada and globally. The ultimate significance of biotechnology is that it involves technologies of science and of power as well as technologies of the human body and life itself. For all these reasons, a fuller range of democratic arenas must be used and new ones deployed in the three bio-realms.

Notes

Introduction

1 The OECD guidance to statistical agencies immediately adds the proviso that this single definition should always be accompanied by the 'list-based' definition, which at present consists of six OECD groupings of biotechnology techniques.
2 These include approaches centred explicitly and formally on concepts such as policy communities, neo-institutionalist analysis, and varieties of political economy.

Chapter 2

1 This chapter focuses on federal government and international biotechnology policy, as does the book as a whole. Aspects of provincial biotechnology policy also emerge and will be discussed in later chapters as further key aspects of multi-level governance and democratic politics.

Chapter 3

1 Data for SFT mentions of biotech are drawn from the authors' reading of Speeches from the Throne and Motions for Address in Reply, various years, http://www.2.parl.gc.ca//Parlinfo/compilations/parliament/ThroneSpeech.aspx?Language=E. Throne Speeches are also quoted and summarized in the various editions of *How Ottawa Spends* for a thirty-year period since 1980 as assessed by the School of Public Policy and Administration at Carleton University.
2 Data below are drawn from a review of Budget Speeches. See Budgets: Historical List from 1867 to Date. www.2.parl.gc.ca//Parlinfo/compilations/parliament/budget.aspx.

Chapter 4

1 In addition to other sources cited, this account draws on interviews and on presentations and discussion at a CBAC workshop held in 2000 and attended by the authors. See Canadian Food Inspection Agency 2000a, 2000b, 2000c; Canadian Food Inspection Agency and Health Canada 2000; and Health Canada 2000a, 2000b, and 2000c.

Chapter 5

1 These data were supplied to the authors in 2004 by the Biotechnology Secretariat of Industry Canada.
2 Calculated by the authors from the Health Canada website data base on biologics, 1994–2009.
3 See http://www.bio.org/speeches/pubs/er/statistics.asp.

Chapter 6

1 For other descriptions of this history, see Manseau (2003), Montpetit (2003), Montpetit, Scala, and Fortier (2004), Scala (2007).
2 The use of these arguments to justify federal intervention is not mutually exclusive. The legislation establishing the Public Health Agency of Canada in 2006 appears to employ the criminal law powers and the peace, order, and good government powers as the basis for that agency (Tiedemann 2006).
3 In fact, in the reference decision, all nine justices seem to accept that the offences provided for in sections 60 and 61 of the AHR Act that relate to provisions are constitutionally valid and thus not in dispute.
4 The 2012 federal budget announced that AHRC as a separate agency would be eliminated, as part of its austerity program but also because of constitutional concerns.

Chapter 7

1 One of us (Prince) appeared as a witness before both a House of Commons committee and a Senate committee that dealt with the Assisted Human Reproduction Act and the creation of the agency.
2 As we have seen in chapter 6, a successful Quebec court challenge contributed to the Harper government's decision in the 2012 budget to eliminate Assisted Human Reproduction Canada.

References

Abdelgafar, Basma, and Halla Thorsteinsdottir. 2007. 'Promoting Partnerships in Biotechnology for Development.' In *Innovation, Science, Environment: Canadian Policies and Performance 2007–2008*, edited by Bruce Doern, 137–61. Montreal and Kingston: McGill-Queen's University Press.

Abergel, Elizabeth, and Katherine Barrett. 2002. 'Putting the Cart before the Horse: A Review of Biotechnology Policy in Canada.' *Journal of Canadian Studies* 37 (3): 135–61.

Abraham, Caroline. 2009. 'Researchers Fear Stagnation under Tories.' *Globe and Mail*, 2 March.

– 2012. 'Reproductive Technology: Unnatural Selection.' *Globe and Mail*, 7 January.

Abraham, J., and G. Lewis. 2000. *Regulating Medicines in Europe: Competition, Expertise and Public Health*. London: Routledge.

– 2003. 'The Europeanization of Medicines Regulation.' In *Regulation of the Pharmaceutical Industry*, edited by J. Abraham and H. Lawton Smith, 42–81. London: Palgrave Macmillan.

Adams, John. 1995. *Risk*. London: UCL.

Agranoff, Robert. 2007. *Managing within Networks: Adding Value to Public Organizations*. Washington, DC: Georgetown University Press.

Allingham-Hawkins, Diane. 2007. 'Addressing Challenges to Providing Medical Genetics Services.' In *Human Genetics Licensing Symposium*, 9–13. Ottawa: Health Canada.

Andrée, Peter. 2009. *Genetically Modified Diplomacy: The Global Politics of Agricultural Biotechnology and the Environment*. Vancouver: University of British Columbia Press.

Appleyard, Bryan. 1999. *Brave New Worlds*. London: Harper Collins.

Assisted Human Reproduction Canada. 2007. *Report on Plans and Priorities*. Ottawa: Assisted Human Reproduction Canada.

Atkinson-Grosjean, Janet. 2006. *Public Science, Private Interests: Culture and Commerce in Canada's Networks of Centres of Excellence*. Toronto: University of Toronto Press.

Aucoin, Peter. 1997. *The New Public Management: Canada in Comparative Perspective*. Montreal and Kingston: McGill-Queen's University Press.

– 2003. 'Independent Foundations, Public Money and Public Accountability: Whither Ministerial Accountability as Democratic Governance?' *Canadian Public Administration* 46 (1): 1–26.

– 2008. 'New Public Management and New Public Governance: Finding the Balance.' In *Professionalism and Public Service: Essays in Honour of Kenneth Kernaghan*, edited by David Siegel and Ken Rasmussen, 16–33. Toronto: University of Toronto Press.

Bache, Ian, and Matthew Flinders, eds. 2004. *Multi-Level Governance*. Oxford: Oxford University Press.

Baldwin, Robert, Colin Scott, and Christopher Hood, eds. 1998. *A Reader on Regulation*. Oxford: Oxford University Press.

Barrett, Katherine, and Elisabeth Abergel. 2000. 'Breeding Familiarity: Environmental Risk Assessment for Genetically Engineered Crops in Canada.' *Science and Public Policy* 27 (1): 2–12.

Bauer, Martin, and George Gaskell. 2002. *Biotechnology: The Making of a Global Controversy*. Cambridge, UK: Cambridge University Press.

Baylis, F., and M. Herder. 2009. 'Policy Design for Human Embryo Research in Canada: A History (Part 1 of 2).' *Bioethical Inquiry* 6: 109–22.

BBC News. 2007. 'Cancer-Drug Refund Scheme Backed.' 4 June. http://news.bbc.co.uk/2/hi/health/6713503.stm.

Bellamy, Richard, and Antonino Palumbo, eds. 2010. *From Government to Governance*. London: Ashgate.

Benkler, Yochai. 2006. *The Wealth of Networks*. New Haven, CT: Yale University Press.

Berger, Peter L., and Thomas Luckmann. 1967. *The Social Construction of Reality: A Treatise on the Sociology of Knowledge*. New York: Anchor Books.

Bettig, Ronald V. 1996. *Copyrighting Culture: The Political Economy of Intellectual Property*. Boulder, CO: Westview.

Bhat, M.G. 1996. 'Trade-Related Intellectual Property Rights to Biological Research: Socioeconomic Implications for Developing Countries.' *Ecological Economics* 19 (3): 205–17.

Bickerton, James, and Alain-G. Gagnon. 2009. *Canadian Politics*. 5th ed. Toronto: University of Toronto Press.

BIOTECanada. 2008. 'Canadians Value Biotech: 2008 Annual Polling Results Overview.' News release, BIOTECanada, September.

– 2009. 'Economic Stimulus Tools to Keep Canada's Research Intensive Emerging Companies Viable.' Ottawa: BIOTECanada.

Blackwell, Tom. 2009. 'The Impotency of Canada's Fertility Laws.' *National Post*, 13 February.

– 2010a. 'Fertility Law Leaves Us in Limbo, Doctors Say.' *National Post*, 30 April.

– 2010b. 'Red Flag Raised at Fertility Agency.' *National Post*, 20 April.

Blank, Robert H., and Samuel M. Hines Jr. 2001. *Biology and Political Science*. London: Routledge.

Bollier, David. 2002. *Silent Theft: The Private Plunder of Our Common Wealth*. New York: Routledge.

Bombard, Yvonne, Gerry Veenstra, Jan M. Friedman, Susan Creighton, Lauren Currie, Jane S. Paulsen, Joan L. Bettor, and Michael R. Hayden. 2009. 'Perceptions of Genetic Discrimination among People at Risk for Huntington's Disease: A Cross-Sectional Survey.' *BMJ Online*. DOI: 10.1136/bmj.b2175.

Borins, Sandford, and David Brown. 2008. 'E-consultation: Technology at the Interface between Civil Society and Government.' In *Professionalism and Public Service: Essays in Honour of Kenneth Kernaghan*, edited by David Siegel and Ken Rasmussen, 178–206. Toronto: University of Toronto Press.

Bouckaert, Geert, Guy Peters, and Koen Verhoest. 2010. *Shifting Patterns of Public Management*. London: Palgrave.

Boyer, Peter J. 2010. 'The Covenant.' *New Yorker*, 6 September.

Brown, Bonnie, Nancy Miller Chenier, and Sonya Norris. 2003. 'Committees as Agents of Public Policy: The Standing Committee on Health.' *Canadian Parliamentary Review* (Autumn 2003): 4–8.

Brunk, Conrad B., and Harold Coward, eds. 2009. *Acceptable Genes: Religious Traditions and Genetically Modified Foods*. Albany: State University of New York Press.

Buchanan, A. 1996. 'Choosing Who Will Be Disabled: Genetic Intervention and the Morality of Inclusion.' *Social Philosophy and Policy* 13 (2): 18–46.

Campbell, R.M., and L.A. Pal. 1989. *The Real Worlds of Canadian Politics: Cases in Process and Policy*. Peterborough, ON: Broadview.

Campbell, R.M., L.A. Pal, and Michael Howlett. 2004. 4th ed. *The Real Worlds of Canadian Politics: Cases in Process and Policy*. Peterborough, ON: Broadview.

Canada. 2005. *Canadian Institutes of Health Research: 2005–2006 Report on Plans and Priorities*. Ottawa: Treasury Board Secretariat.

- 2009a. *BioBasics.* http://www.biobasics.gc.ca/english/view.asp?x=556.
- 2009b. *Final Report of the Independent Investigator into the 2008 Listeriosis Outbreak* (the Weatherill Report). Ottawa: Agriculture and Agri-Food Canada.

Canada Foundation for Innovation. 2010. *Innovation, Made in Canada, Best in the World.* Annual Report 2009–2010. Ottawa: Canada Foundation for Innovation.

Canada's Research-Based Pharmaceutical Companies. 2005. 'CDR Report: Patients Fail to Gain Timely Access to New Medicines.' News release, 14 October.
- 2006. 'Study Shows Canada's Common Drug Review Is a Barrier to Access to New Medicines for Patients.' News release, 8 December.

Canadian Agency for Drugs and Technologies in Health. 2006. *Common Drug Review.* http://cadth.ca/index.php/en/cdr.
- 2007. 'Submission Brief to House of Commons Committee on Health.' 25 April.
- 2008. *Annual Report.* Ottawa: Canadian Agency for Drugs and Technologies in Health.

Canadian Biotechnology Advisory Committee. 2002. *The Regulation of Genetically Modified Foods.* Ottawa: Canadian Biotechnology Advisory Committee.
- 2006. *Human Genetic Materials, Intellectual Property and the Health Sector.* Ottawa: Canadian Biotechnology Advisory Committee.

Canadian Biotechnology Secretariat. 2002. *Canadian Biotechnology Strategy Overall Performance Report 1999–2002.* Ottawa: Canadian Biotechnology Strategy.
- 2004. 'The Government of Canada Blueprint for Biotechnology: Realizing Canada's Potential.' Biotechnology Assistant Deputy Ministers' Coordinating Committee. Ottawa: Canadian Biotechnology Strategy.

Canadian Coalition for Genetic Fairness. 2010. 'About CCGF.' http://www.ccgf-cceg.ca/en/about-ccgf.

Canadian Food Inspection Agency. 2000a. Presentation by Catherine Italiano on Feeds Act and Regulations to Canadian Biotechnology Advisory Committee. *GM Food Project Steering Committee: Regulators Workshop Technical Briefing Decks.* Ottawa: Canadian Biotechnology Advisory Committee, 23 June.
- 2000b. Presentation by Phil MacDonald on Confined Release Field Trials to Canadian Biotechnology Advisory Committee. *GM Food Project Steering Committee: Regulators Workshop Technical Briefing Decks.* Ottawa: Canadian Biotechnology Advisory Committee, 23 June.

– 2000c. Presentation by Stephen Yarrow on Harmonization of Regulatory Systems to Canadian Biotechnology Advisory Committee. *GM Food Project Steering Committee: Regulators Workshop Technical Briefing Decks.* Ottawa: Canadian Biotechnology Advisory Committee, 23 June.

Canadian Food Inspection Agency and Health Canada. 2000. Presentation by Bart Bilmer, Tim Flaherty, and Claudette Dalpe on Overview of Biotechnology Regulation in the Federal Government to Canadian Biotechnology Advisory Committee. *GM Food Project Steering Committee: Regulators Workshop Technical Briefing Decks.* Ottawa: Canadian Biotechnology Advisory Committee, 23 June.

– 2009a. *Plant Biosafety.* http://www.inspection.gc.ca/english/plaveg/bio.

– 2009b. *Regulating Agricultural Biotechnology.* http://www.inspection.gc.ca/english/animal/biotech.

– 2009c. Summary Tables of Confined Field Trials (Various Years). http://www.inspection.gc.ca/english/plaveg/bio.

Carlson, Kathryn B. 2010. 'Baby by Stealth: Reproduction Law Forcing "Dangerous Alternatives."' *National Post,* 12 March.

Carolan, Michael S. 2010. 'The Mutability of Biotechnology Patents: From Unwieldy Products of Nature to Independent "Objects."' *Theory, Culture and Society* 27: 110–29.

Carpenter, Daniel. 2010. *Reputation and Power: Organizational Image and Pharmaceutical Regulation at the FDA.* Princeton: Princeton University Press.

Castle, David, ed. 2009. *The Role of Intellectual Property Rights in Biotechnology Innovation.* Cheltenham, UK: Edward Elgar.

Caulfield, Timothy. 1996. 'The Commercialization of Human Genetics: A Discussion of Issues.' OCA Research Paper. Ottawa: Industry Canada.

– 2003. 'Human Cloning Laws, Human Dignity and the Poverty of Policy-Making Dialogue.' *BMC Medical Ethics* 4 (3): 1–7.

– 2009a. 'Biotechnology Patents, Public Trust and Patent Pools: The Need for Governance.' In *The Role of Intellectual Property Rights in Biotechnology Innovation,* edited by David Castle, 357–68. Cheltenham, UK: Edward Elgar.

– 2009b. 'The Commercialization of Genomic Research in Canada.' Presentation to Genome Quebec Conference on the Commercialization of Genomic Research in Canada, University of Montreal, 30 January.

Caulfield, Timothy, and T. Bubela. 2007. 'Why a Criminal Ban? Analyzing the Arguments against Somatic Cell Nuclear Transfer in the Canadian Parliamentary Debate.' *American Journal of Bioethics* 7 (2): 51–61.

Caulfield, Timothy, N. Ries, P. Ray, C. Schuman, and B. Wilson. 2010. 'Direct-to-Consumer Genetic Testing: Good, Bad or Benign?' *Clinical Genetics* 77: 101–5.

Coleman, William, and Grace Skogstad, eds. 1990. *Policy Communities and Public Policy in Canada*. Toronto: Copp Clark Pitman.

Collins, Francis S. 2010. *The Language of Life: DNA and the Revolution in Personalized Medicine*. New York: Harper Collins.

Commission of Inquiry on the Blood System of Canada (The Krever Inquiry). 1997. *Final Report*. Ottawa: Minister of Public Works and Government Services.

Conference Board of Canada. 2004. *Understanding Health Care Cost Drivers and Escalators*. Ottawa: Conference Board of Canada.

– 2010a. *Intellectual Property in the 21st Century*. Ottawa: Conference Board of Canada.

– 2010b. *Conflicting Forces for Canadian Prosperity: Examining the Interplay between Regulation and Innovation*. Ottawa: Conference Board of Canada.

Connelly, James, and Graham Smith. 1999. *Politics and the Environment*. London: Routledge.

Connor, Steve. 2010a. 'Autism and Genetics: A Breakthrough That Sheds Light on a Medical Mystery.' *Independent*, 10 June.

Connor, Steve. 2010b. 'It Pays Not to Cultivate GM Crops, Survey Finds.' *Independent*, 8 October.

Cook-Deegan, Robert. 2009. 'Aligning Incentives between Saving Lives and Making Money.' Presentation to Genome Quebec Conference on the Commercialization of Genomic Research in Canada, University of Montreal, 30 January.

Council of Canadians and Greenpeace. 2001. 'Canadian Organizations Decry Government Biotech Policies.' News release, 4 April.

Cruikshank, Barbara. 1999. *The Will to Empower: Democratic Citizens and Other Subjects*. Ithaca, NY: Cornell University Press.

de Beer, Jeremy, and Mauricio Guaranga. 2011. *Intellectual Property Management: Issues and Options*. Policy brief no. 4. Ottawa: Genome Canada.

Decima Research. 2006. *Public Engagement on the Future Government of Canada Role in Biotechnology*. Report prepared for Canadian Biotechnology Secretariat, Industry Canada, June.

Deckha, M. 2009. 'Holding onto Humanity: Animals, Dignity, and Anxiety in Canada's *Assisted Human Reproduction Act*.' *Unbound* 5: 21–54.

Dewing, Michael. 2008. 'Parliamentary Committee Review of Regulations.' http://www2.parl.gc.ca/content/lop/researchpublications/prb0563-e.htm.

Dickson, Barney, and Rosie Cooney. 2005. *Biodiversity and the Precautionary Principle: Risk and Uncertainty*. London: Earthscan.

Doern, Bruce. 1972. *Science and Politics in Canada*. Montreal and Kingston: McGill-Queen's University Press.

- 1981. *The Peripheral Nature of Scientific and Technological Controversy in Federal Policy Formation.* Ottawa: Science Council of Canada.
- 1996. 'Looking for the Core: Industry Canada and Program Review.' In *How Ottawa Spends: 1996–97: Life under the Knife,* edited by Gene Swimmer, 73–98. Ottawa: Carleton University Press.
- 1999. *Global Change and Intellectual Property Agencies.* London: Pinter.
- 2000. 'Biotechnology, Public Confidence and Governance.' Paper for the Workshop on Public Confidence in Biotechnology, Industry Canada, Ottawa, 4 March.
- 2002a. 'The Chrétien Liberals' Third Mandate.' In Doern 2002b, 1–19.
- ed. 2002b. *How Ottawa Spends 2002–2003: The Security Aftermath and National Priorities.* Toronto: Oxford University Press.
- 2003. *Regulatory Regimes for the Safety and Efficacy of Biotechnological Health Products: Changing Pressures, Products and Processes.* Paper prepared for the Canadian Biotechnology Advisory Committee.
- 2004. *The Agri-Food Sector and Federal Policies and Priorities: A Public Policy Framework Discussion Framework.* Paper prepared for Agriculture and Agri-Food Canada.
- 2007. *Red Tape, Red Flags: Regulation in the Innovation Age.* Ottawa: Conference Board of Canada.
- 2009. 'The Granting Councils and the Research Granting Process: Core Values in Federal Government–University Interactions.' In *Research and Innovation Policy: Changing Federal Government–University Relations,* edited by Bruce Doern and Chris Stoney, 89–123. Toronto: University of Toronto Press.
- 2010. *The Governance and Reform of Food Safety Systems: Canada in a Comparative Context.* Study prepared for Agriculture and Agri-Food Canada.
Doern, Bruce, and Thomas Conway. 1994. *The Greening of Canada: Federal Institutions and Decisions.* Toronto: University of Toronto Press.
Doern, Bruce, and Monica Gattinger. 2003. *Power Switch: Energy Regulatory Governance in the 21st Century.* Toronto: University of Toronto Press.
Doern, Bruce, Margaret M. Hill, Michael J. Prince, and Richard J. Schultz, eds. 1999. *Changing the Rules: Canadian Regulatory Regimes and Institutions.* Toronto: University of Toronto Press.
Doern, Bruce, and Robert Johnson, eds. 2006. *Rules, Rules, Rules, Rules: Multilevel Regulatory Governance.* Toronto: University of Toronto Press.
Doern, Bruce, and Jeffrey Kinder. 2007. *Strategic Science in the Public Interest: Canada's Government Laboratories and Science-Based Agencies.* Toronto: University of Toronto Press.

Doern, Bruce, and Richard Levesque. 2002. *The National Research Council of Canada in the Innovation Policy Era: Changing Hierarchies, Networks and Markets*. Toronto: University of Toronto Press.

Doern, Bruce, Allan Maslove, and Michael J. Prince. 1988. *Budgeting in Canada: Politics, Economics and Management*. Ottawa: Carleton University Press.

Doern, Bruce, and Richard W. Phidd. 1992. *Canadian Public Policy: Ideas, Structure, Process*. 2nd ed. Toronto: Nelson Canada.

Doern, Bruce, and Peter W.B. Phillips. 2012. *The Genomics 'Regulatory-Science' Regime: Issues and Options*. Policy brief no. 5. Ottawa: Genome Canada.

Doern, Bruce, and Ted Reed, eds. 2000. *Risky Business: Canada's Changing Science-Based Policy and Regulatory Regime*. Toronto: University of Toronto Press.

Doern, Bruce, and Markus Sharaput. 2000. *Canadian Intellectual Property: The Politics of Innovating Institutions and Interests*. Toronto: University of Toronto Press.

Doern, Bruce, and Heather Sheehy. 1999. 'The Federal Biotechnology Regulatory System: A Commentary on an Institutional Work in Progress.' In *Biotechnology and the Consumer*, edited by B.M. Knoppers and Alan D. Mathios, 56–71. Dordrecht, Netherlands: Kluwer, 1999.

Doern, Bruce, and Christopher Stoney. 2009. *Research and Innovation Policy: Changing Federal Government–University Relations*. Toronto: University of Toronto Press.

– 2010. 'Double Deficit: Fiscal and Democratic Challenges in the Harper Era.' In *How Ottawa Spends 2010–2011: Recession, Realignment and the New Deficit Era*, edited by Bruce Doern and Christopher Stoney, 3–30. Montreal and Kingston: McGill-Queen's University Press.

Doern, Bruce, and Stephen Wilks. 2007. 'Accountability and Multi-Level Governance in UK Regulation.' In *Regulatory Review 2006–2007*, edited by Peter Vass, 341–72. Bath, UK: Centre for the Study of Regulated Industries, University of Bath.

Doremus, P.N. 1996. 'The Externalization of Domestic Regulation: Intellectual Property Rights Reform in a Global Era.' *Science Communication* 17 (2): 137–62.

Drahos, P. 1996. 'Global Law Reform and Rent-Seeking: The Case of Intellectual Property.' *Australian Journal of Corporate Law* 7: 45–61.

– 1997. 'Thinking Strategically about Intellectual Property Rights.' *Telecommunications Policy* 21 (3): 201–11.

Drahos, P., and John Braithwaite. 2002. *Information Feudalism: Who Owns the Knowledge Economy?* London: Earthscan.

Drahos, P., and R. Mayne. 2002. *Global Intellectual Property Rights: Knowledge, Access and Development*. London: Palgrave.

Dryzek, John, and Patrick Dunleavy. 2009. *Theories of the Democratic State.* London: Palgrave Macmillan.

Dutfield, Graham. 2003. *Intellectual Property Rights and the Life Science Industries.* London: Ashgate.

– 2008. 'Turning Plant Varieties into Intellectual Property: The UPOV Convention.' In *The Future Control of Food,* edited by G. Tansey and T. Rajotte, 27–47. London: Earthscan.

Easy DNA. 2010. DNA Testing Knowledge Base. http://www.easy-dna.com/dnanews.

Economist. 2007a. 'Pharmaceuticals: A Gathering Storm.' 9 June.

– 2007b. 'Pharmaceuticals: Beyond the Blockbuster.' 30 June.

– 2009. 'Medicine Goes Digital: The Convergence of Biology and Engineering.' Special report, 18 April.

– 2011. 'Rows over GM Crops: Seed of Change.' 8 January.

Egan, Timothy. 2011 'Frankenfish Phobia.' *New York Times,* 17 March.

Einseidel, Edna F., and Frank Timmermans. 2005. *Crossing Over: Genomics in the Public Arena.* Calgary: University of Calgary Press.

Environment Canada. 2000. 'Backgrounder on the Biosafety Protocol to the UN Convention on Biological Diversity.' Ottawa: Environment Canada.

Etzioni, Amitai. 1968. *The Active Society.* New York: Free Press.

European Group on Ethics. 2002. 'Ethical Aspects of Patenting Inventions Involving Human Stem Cells.' http://ec.europa.eu/european_group_ethics/docs/avis16_en.pdf.

Evans, Barbara J. 2009. 'Seven Pillars of a New Evidentiary Paradigm: The Food, Drug and Cosmetic Act Enters the Genomic Era.' *Notre Dame Law Review* 85: 22–36.

External Advisory Committee on Smart Regulation. 2004. *Smart Regulation: A Regulatory Strategy for Canada.* Report to the Government of Canada. September.

Fitzpatrick, T. 2001. 'Before the Cradle: New Genetics, Biopolicy and Regulated Eugenics.' *Journal of Social Policy* 30 (4): 589–612.

Flinders, Matthew. 2001. *The Politics of Accountability in the Modern State.* London: Ashgate.

– 2008. *Delegated Governance and the British State: Walking without Order.* Oxford: Oxford University Press.

Forge, Frederic. 2007. *Biofuels: An Energy, Environmental or Agricultural Policy?* Ottawa: Library of Parliament.

Foucault, M. 1980. *Power/Knowledge.* Brighton: Harvester.

– 2008. *The Birth of Biopolitics: Lectures at the College de France, 1978–1979.* London: Palgrave Macmillan.

Fox, Dov. 2008. 'The Regulation of Biotechnologies: Four Recommendations.' *Perspectives* 38 (2): 111–26.

Fukuyama, F. 2002. *Our Posthuman Future: Consequences of the Biotechnology Revolution.* New York: Farrar, Straus and Giroux.

Fukuyama, Frances, and Caroline Wagner. 2001. 'Governance Challenges of Technological Revolutions.' In *Science, Technology and Governance*, edited by John de la Mothe, 188–209. London: Continuum.

Gelineault, Caroline. 2002. 'Genome Canada at a Glance.' Unpublished research paper, School of Public Policy and Administration, Carleton University, September.

Genome Canada. 1999. 'Genome Canada Business Plan: Discussion Paper.' Ottawa: Genome Canada, November.

– 2009a. 'About GE³LS.' http://www.genomecanada.ca/en/ge3ls.

– 2009b. 'About Genome Canada.' http://www.genomecanada.ca/en/about.

Genome Quebec. 2009. 'Genomics Issues.' http://www.genomequebec.com/v2009/ethique/index.

Grace, Eric. 1997. *Biotechnology Unzipped.* Washington: National Academy.

Gray, C.H. 2002. *Cyborg Citizens: Politics in the Posthuman Age.* New York: Routledge.

Greenpeace. 2008. 'Parliament Denies Canadians Mandatory Labelling of GE Foods.' (Greenpeace) http://www.greenpeace.org/Canada/en/recent.

Hale, Geoffrey. 2002. 'Innovation and Inclusion: Budgetary Policy, the Skills Agenda, and the Politics of the New Economy.' In Doern 2002b, 20–47.

Hancher, Leigh, and Michael Moran, eds. 1989. *Capitalism, Culture and Regulation.* Oxford: Clarendon.

Haraway, D. 1989. 'The Biopolitics of Postmodern Bodies: Determinations of Self in Immune System Discourse.' *Differences: A Journal of Feminist Cultural Studies* 1 (1): 3–43.

Harmsen, Eef, Rob Sladek, and Andrew Orr. 2006. 'Functional Genomics and Proteomics in Personalized Medicine.' *Bioscienceworld.* http://bioscienceworld.ca/FunctionalGenomicsandProteomicsinPersonalizedMedicine21stCenturyApproachestoComplexDiseases.

Harris, John. 2010. 'Promise and Risks from Life Not as We Know It.' *Financial Times,* 27 May.

Hauskeller, Michael. 2007. *Biotechnology and the Integrity of Life: Taking Public Fears Seriously.* London: Ashgate.

Health Canada. 2000a. Presentation by Karen McIntyre on Case Study: Novel Food Safety Assessment for a Genetically Modified Canola to Canadian Biotechnology Advisory Committee. *GM Food Project Steering Committee: Regulators Workshop Technical Briefing Decks.* Ottawa: CBAC, 23 June.

- 2000b. Presentation by Paul Mayers on Food Safety Aspects of Genetically Modified Plants: Key Issues, International Perspective to Canadian Biotechnology Advisory Committee. *GM Food Project Steering Committee: Regulators Workshop Technical Briefing Decks*. Ottawa: CBAC, 23 June.
- 2000c. Presentation by William Yan on the Regulatory Approval Process of Genetically Modified Organisms in Canada: Overview from Health Canada to Canadian Biotechnology Advisory Committee. *GM Food Project Steering Committee: Regulators Workshop Technical Briefing Decks*. Ottawa: CBAC, 23 June.
- 2006a. *Blueprint for Renewal: Transforming Canada's Approach to Regulating Health Products and Food*. Ottawa: Health Canada.
- 2006b. 'Canada's Access to Medicine's Regime: Consultation Paper.' Ottawa: Health Canada.
- 2007a. *Blueprint for Renewal II: Modernizing Canada's Regulatory System for Health Products and Food*. Ottawa: Health Canada.
- 2007b. 'Decision-Making Framework Status Paper, May 2007.' Draft paper. Ottawa: Health Canada.
- 2009a. 'Biologics, Radiopharmaceuticals and Genetic Therapies.' Ottawa: Health Canada.
- 2009b. 'Departmental Biotechnology Office.' Ottawa: Health Canada. http://www.hc-sc.gc.ca/sr-sr/pubs/biotech/over-sur_bio-eng.
- 2009c. *Drug Licensing Process*. Ottawa: Health Canada.
- 2009d. 'What Are Novel Foods and Genetically Modified (GM) Foods?' Ottawa: Health Canada. http://www.hc-sc.gc.ca/fn-an/gmf-agm/index-eng.php.
- 2009e. 'Post-Market Surveillance of Drug Products Derived from Biotechnology.' Ottawa: Health Canada. http://www.hc-sc.gc.ca/sr-srpubs/biotech/.
- 2009f. 'The Progressive Licensing Framework: Concept Paper for Discussion.' Ottawa: Health Canada. http://www.hc-sc.gc.ca/dhp-mps/homologation-licensing/development.
- 2010a. 'Canada's Biotechnology Strategy.' Ottawa: Health Canada. http://www.hc-sc.gc.ca/sr-sr/biotech/role/strateg-eng.php.
- 2010b. *Keeping an Eye on Prescription Drugs, Keeping Canadians Safe*. Ottawa: Health Council of Canada.
- 2010c. 'Publication of Proposed Assisted Human Reproduction Regulations Delayed until Supreme Court Appeal Is Decided.' Ottawa: Health Canada. http://www.hc-sc.gc.ca/hl-vs/reprod/hc-sc/legislation/delay-interruption-eng.php.
Hebert, Monique. 2004. 'A Review of Bill C-6: Assisted Human Reproduction Act.' Library of Parliament.

Hickman, Martin. 2009. 'Big Stores Counting the Cost of Ban on GM Food.' *Independent*, 1 September.

Hill, Margaret. 1999. 'Managing the Regulatory State: From "Up" to "In and Down" to "Out and Across."' In Doern, Hill, Prince, and Schultz.

Hochedlinger, Konrad. 2010. 'Your Inner Healers.' *Scientific American*, May, 29–36.

Hood, Christopher. 1998. *The Art of the State*. Oxford: Oxford University Press.

House of Commons Standing Committee on Agriculture and Agri-Food. 1998. *Capturing the Advantage: Agriculture and Agri-Food*. Ottawa: House of Commons Standing Committee on Agriculture and Agri-Food, May.

House of Commons Standing Committee on Environment and Sustainable Development. 1996. *Biotechnology Regulation in Canada: A Matter of Confidence*. Ottawa: House of Commons Standing Committee on Environment and Sustainable Development.

Howlett, Michael. 2002. 'Do Networks Matter? Linking Policy Network Structure to Policy Outcomes: Evidence from Four Canadian Policy Sectors 1990–2000.' *Canadian Journal of Political Science* 35 (2): 235–67.

Howlett, Michael, and M. Ramesh. 2003. *Studying Public Policy*. 2nd ed. Toronto: Oxford University Press.

Hubbard, Ruth, and Gilles Paquet. 2007. 'Public-Private Partnerships: P3 and the "Porcupine Problem."' In *How Ottawa Spends 2007–2008: The Harper Conservatives – Climate of Change*, edited by Bruce Doern, 254–72. Montreal and Kingston: McGill-Queen's University Press.

Hughes, B. 2009. 'Disability Activisms: Social Model Stalwarts and Biological Citizens.' *Disability & Society* 24 (6): 677–88.

Humphries, Courtney. 2010. 'FDA Takes on Personalized Medicine.' *Technology Review*, March/April, 12–14.

Hunt, Wayne A. 1999. 'Genetically Modified Politics.' *Policy Options*, November, 59–62.

Hutchings, J.A., C.J. Walters, and R.L. Haedrich. 1997. 'Is Scientific Inquiry Incompatible with Government Information Control?' *Canadian Journal of Fisheries and Aquatic Sciences* 54: 1198–210.

Independent. 2000. 'The Blair-Clinton Statement on the Human Genome.' 15 March.

Industry Canada. 1998. *Renewal of the Canadian Biotechnology Strategy: International Issues Report*. Ottawa: Industry Canada.

– 2002. *Achieving Excellence: Canada's Innovation Strategy*. Ottawa: Industry Canada.

– 2005. 'Government of Canada Federal Stewardship Framework for Biotechnology.' Draft discussion paper, July.

- 2007. *Mobilizing Science and Technology to Canada's Advantage*. Ottawa: Industry Canada.
- 2009a. 'Life Sciences Gateway, Biotechnology Sector Overview.' Ottawa: Industry Canada. http://www.ic.gc.ca/eic/site/lsg-pdsv.
- 2009b. *Progress Report on Science, Technology and Innovation Council*. Ottawa: Industry Canada.

Institute of Genetics. 2009. *About IG*. http://www.cihr-irsc.gc.ca/e/ig-aboutig.

International Expert Group on Biotechnology, Innovation and Intellectual Property. 2008. *Toward a New Era of Intellectual Property: From Confrontation to Negotiation*. Montreal: McGill Centre on Intellectual Property Policy, McGill University.

Jack, Andrew. 2009. 'Remedy for a Malady.' *Financial Times*, 15 August.
- 2010. 'Big Pharma Aims for Reinvention.' *Financial Times*, 13 May.

Jackman, Martha. 2000. 'Constitutional Jurisdiction over Health Care in Canada.' *Health Law Journal* 8: 95–117.

Jarvis, Bill. 2000. 'A Question of Balance: New Approaches for Science-Based Regulation.' In Doern and Reed, 307–33.

Jones, David P., and Anne S. de Villars. 1999. *Principles of Administrative Law*. 3rd ed. Toronto: Carswell.

Jones, M., and B. Salter. 2009. 'Proceeding Carefully: Assisted Human Reproduction Policy in Canada.' *Public Understanding of Science* 9 (2): 1–15.

Jordan, Andrew, and T. O'Riordan. 1995. 'The Precautionary Principle in UK Environmental Policy Making.' In *UK Environmental Policy in the 1990s*, edited by Tim S. Gray, 57–84. London: Macmillan.

Judson, Olivia. 2010a. 'Baby Steps to New Life Forms.' *New York Times*, 27 May.
- 2010b. 'The Human Genome Project.' *New York Times*, 8 June.

Kanter, James. 2010. 'EU Clears Biotech Potato for Cultivation.' *New York Times*, 3 March. http://www.nytimes.com/2010/06/26/business/global/03potato.html.

Kinder, Jeffrey S. 2010. 'Government Laboratories: Institutional Variety, Change and Design Space.' PhD diss., Carleton University.

Klar, Estee. 2010. Letter to the editor, *Globe and Mail*, 12 June.

Knoppers, Bartha M., ed. 1998. *Socio-ethical Issues in Human Genetics*. Cowansville, QC: Les Éditions Yvon Blais.

Knoppers, Bartha M., and Rosario M. Isasi. 2004. 'Regulatory Approaches to Reproductive Genetic Testing.' *Human Reproduction* 19 (12): 2695–701.

Knoppers, Bartha M., and Alan Mathios, eds. 1998. *Biotechnology and the Consumer*. Dordrecht: Kluwer.

Knowles, Lori P. 2010. 'Religion and Stem Cell Research.' Stem Cell Network. http://www.stemcellnetwork.ca/index.php.

Kolata, Gina. 2010. 'Stem Cell Biology and Its Complications.' *New York Times*, 24 August. http://www.nytimes.com/2010/08/25/health/research/25cell.html.

KPMG. 2007. *Evaluation of Foundations*. Toronto: KPMG.

– 2009. *Evaluation of Genome Canada: Final Report*. Toronto: KPMG.

Kumar, Nikil. 2010. 'Biotech Firms Hit as Cash Dries Up and Research Shifts Eastward.' *Independent*, 8 November.

Kuyek, Devlin. 2002. *The Real Board of Directors: The Construction of Biotechnology Policy in Canada, 1980–2002*. Sorrento, BC: Ram's Horn.

Kuzma, Jennifer, and Todd Tanji. 2010. 'Unpackaging Synthetic Biology: Identification of Oversight Policy Problems and Options.' *Regulation and Governance* 4 (1): 92–112.

Lal Das, Bhagirath. 1999. *The World Trade Organization: A Guide to the Framework for International Trade*. London: Zed Books.

Landry, Rejean. 2009. 'Reflections on the Commercialization of Academic Genomic Research in Canada.' Presentation to Genome Quebec Conference on the Commercialization of Genomic Research in Canada, University of Montreal, 30 January.

Laucius, Joanne. 2007. 'Patients Buy Cancer Drugs at New Clinic.' *Ottawa Citizen*, 3 June.

Leeder, Jessica. 2010. 'Engineered-in-Canada Salmon Declared Fit for the Dinner Plate.' *Globe and Mail*, 4 September.

Leiss, William. 2000. 'Between Expertise and Bureaucracy: Risk Management Trapped at the Science-Policy Interface.' In Doern and Reed, 49–74.

Leiss, William, and Christina Chociolko. 1995. *Risk and Responsibility*. Montreal and Kingston: McGill-Queen's University Press.

Lemmens, Trudo, and Ron A. Bouchard. 2007. 'The Regulation of Pharmaceuticals in Canada.' In *Canadian Health Law and Policy*, ed. J. Downie, T. Caulfield, and C. Flood. 3rd ed., 121–36. Toronto: Butterworths.

Lemmens, Trudo, Daryl Pullman, and Rebecca Rodal. 2010. *Revisiting Genetic Discrimination Issues in 2010: Policy Options for Canada*. Policy brief no. 2. Ottawa: Genome Canada.

Liagouras, George. 2005. 'The Political Economy of Post-Industrial Capitalism.' *Thesis Eleven* 81: 20, 35.

Locke, Stephen. 1998. 'Modelling the Consumer Interest.' In *Changing Regulatory Institutions in Britain and North America*, ed. Bruce Doern and Stephen Wilks, 162–86. Toronto: University of Toronto Press.

Lopreite, Debora, and Joan Murphy. 2009. 'The Canada Foundation for Innovation as Patron and Regulator.' In Doern and Stoney, 123–47.

Lynas, Mark. 1999. 'The World Trade Organization and GMOs.' *Consumer Policy Review* 9 (6): 214–19.

MacDonald, Mark. 2000. 'Socio-economic versus Science-Based Regulation: Informal Influences on the Formal Regulation of rbST in Canada.' In Doern and Reed, 156–84.

Maeder, Thomas. 2003. 'The Orphan Drug Backlash.' *Scientific American* 288 (5): 70–7.

Maheu, Louis, and Roderick A. Macdonald, eds. 2010. *Challenging Genetic Determinism: New Perspectives on the Gene in Its Multiple Environments.* Montreal and Kingston: McGill-Queen's University Press.

Manseau, F. 2003. 'Canada's Proposal for Legislation on Assisted Human Reproduction.' In *The Regulation of Assisted Reproductive Technology*, ed. J. Gunning and H. Szoke, 45–54. Aldershot, UK: Ashgate.

Marcellin, Sherry S. 2010. *The Political Economy of Pharmaceutical Patents.* London: Ashgate.

Maslove, Allan. 2005. 'Health and Federal–Provincial Fiscal Arrangements: Lost Opportunity.' In *How Ottawa Spends 2005–2006: Managing the Minority*, ed. Bruce Doern, 23–40. Montreal and Kingston: McGill-Queen's University Press.

May, Christopher. 2009. 'On the Border: Biotechnology, the Scope of Intellectual Property and the Dissemination of Scientific Benefits.' In *The Role of Intellectual Property Rights in Biotechnology Innovation*, ed. David Castle, 252–73. Cheltenham, UK: Edward Elgar.

May, Peter J. 2007. 'Regulatory Regimes and Accountability.' *Regulation and Governance* 1 (1): 8–26.

McGillivray, Barbara. 2007. 'Balancing Industry and Health System Interests in Genetics Research and IP.' In *Human Genetics Licensing Symposium*, 44–50. Ottawa: Health Canada.

McGreal, Chris. 2010. 'GM Salmon May Go on Sale in U.S. after Public Consultation.' *Guardian*, 24 August.

McHughen, Alan. 2002. *Biotechnology and Food for Canadians.* Vancouver: Fraser Institute.

McLaren, Margaret A. 2002. *Feminism, Foucault and Embodied Subjectivity.* Albany, NY: State University of New York Press.

Mehta, Michael. 2009. *Biotechnology Unglued: Science, Society and Social Cohesion.* Vancouver: University of British Columbia Press.

Meslin, Eric M. 2010. 'What to Expect from the New Bioethics Commission.' Science Progress. http://www.scienceprogress.org/2010/05/problemsolvers.

Middleton, Benet. 1998. 'Consumerism: A Pragmatic Ideology.' *Consumer Policy Review* 8 (6): 213–17.

Midgley, Mary. 2010. *The Solitary Self: Darwin and the Selfish Gene.* Durham, UK: Acumen Publishing.

Mihlar, Fazil. 1999. 'The Federal Government and the RIAS Process: Origins,
    Need, and Non-Compliance.' In Doern, Hill, Prince, and Schultz, 277–92.
Miller Chenier, Nancy. 1994. *Reproductive Technologies: Royal Commission Final
    Report*. Ottawa: Library of Parliament, Political and Social Affairs Division.
– 2002. *Intergovernmental Consultations on Health: Toward a National Framework
    on Reproductive Technologies*. Ottawa: Library of Parliament. http://dsp-psd.
    pwgsc.gc.ca/Collection-R/LoPBdP/PRB-e/PRB0234-e.pdf.
Mills, Lisa. 2002. *Science and Social Context: The Regulation of Recombinant
    Bovine Growth Hormone in North America*. Montreal and Kingston: McGill-
    Queen's University Press.
Mills, Lisa, and Ashley Weber. 2006. 'Access to Medicines: How Canada
    Amends the Patent Act.' In *How Ottawa Spends 2006–2007: In from the Cold –
    The Tory Rise and the Liberal Demise*, ed. Bruce Doern, 229–46. Montreal and
    Kingston: McGill-Queen's University Press.
Mills, Oliver. 2010. *Biotechnological Inventions: Moral Restraints and Patent Law*.
    Rev. ed. Aldershot, UK: Ashgate.
Millstone, Eric, Eric Brunner, and Sue Mayer. 1999. 'Beyond the "Substantial
    Equivalence" of GM Foods.' *Nature* 401: 525–6.
Ministry of State for Science and Technology. 1980. *Biotechnology in Canada*.
    Ottawa: Ministry of State for Science and Technology.
Mironesco, Christine. 1998. 'Parliamentary Technology Assessment of
    Biotechnologies: A Review of Major TA Reports in the European Union
    and the USA.' *Science and Public Policy* 24 (5): 327–42.
Mitchell, Donald. 2008. 'A Note on Rising Food Prices.' World Bank, Policy
    Research Working Paper 4682, July.
Montpetit, E. 2003. 'Public Consultations in Policy Network Environments:
    The Case of Assisted Human Technology Policy in Canada.' *Canadian
    Public Policy* 29 (1): 95–110.
– 2004. 'Policy Networks, Federalism and Managerial Ideas: How ART Non-
    Decision in Canada Safeguards the Autonomy of the Medical Profession.'
    In *Comparative Biomedical Policy: Governing Assisted Reproductive Technologies*,
    ed. I. Bleiklie, M. Goggin, and C. Rothmayr, 64–81. London: Routledge.
– 2009. 'Has the European Union Made Europe More or Less Democratic?
    Elections, Network Deliberations and Advocacy Groups.' Paper presented
    to the Conference on Bringing Civil Society In: The European Union and
    the Rise of Representative Government, European University Institute,
    Florence, 13 March.
Montpetit, E., F. Scala, and I. Fortier. 2004. 'The Paradox of Deliberative
    Democracy: The National Action Committee on the Status of Women and
    Canada's Policy on Reproductive Policy.' *Policy Sciences* 37: 137–57.

Mooney, Chris. 2005. *The Republican War on Science*. New York: Basic Books.

Morris, S.G. 2007. 'Canada's Assisted Human Reproduction Act: A Chimera of Religion and Politics.' *American Journal of Bioethics* 7 (2): 69–70.

Mouffe, Chantal. 1993. *The Return of the Political*. New York: Verso Books.

Murphy, Joan. 2006. 'Multilevel Regulatory Governance in the Health Sector.' In *Rules, Rules, Rules, Rules: Multilevel Regulatory Governance*, ed. Bruce Doern and Robert Johnson, 305–24. Toronto: University of Toronto Press.

– 2007. 'Transforming Health Sciences Research: From the Medical Research Council to the Canadian Institutes of Health Research.' In *Innovation, Science and Environment: Canadian Policies and Performance 2007–2008*, ed. Bruce Doern, 240–61. Montreal and Kingston: McGill-Queen's University Press.

Mykitiuk, R., J. Nisker, and R. Bluhm. 2007. 'The Canadian Assisted Human Reproduction Act: Protecting Women's Health While Potentially Allowing Human Somatic Cell Transfer into Non-Human Oocytes.' *American Journal of Bioethics* 7 (2): 71–3.

National Biotechnology Advisory Committee. 1998. *Leading into the Next Millennium: Sixth Report*. Ottawa: Industry Canada. http://srrategis.ic.gc.ca?SSG/bh00234e.html.

National Research Council Canada. 1999. *Biotechnology Research Institute: 1998–1999; Performance Report*. Ottawa: National Research Council.

– 2010a. 'Genomics and Health Initiative.' http://www.nrc-cnrc.gc.ca/eng/ibp/ghi.html.

– 2010b. 'Introduction to the Plant Biotechnology Institute.' http://www.pbi.nrc.ca/ENGLISH/introduction-to-pbi.htm.

National Research Council (US). 2010. *The Impact of Genetically Modified Crops on Farm Sustainability in the United States*. Washington: National Research Council, National Academies.

Natural Sciences and Engineering Research Council. 2008. *NSERC Annual Report 2007–2008*. Ottawa: Natural Science and Engineering Research Council.

Networks of Centres of Excellence. 2010. 'Networks and Centres.' http://www.nce-rce.gc.ca/NetworksCentres-CentresReseaux/Index_eng.asp.

New Democratic Party. 2010a. 'Historic NDP GMO Bill Passes Crucial Vote.' News release, 15 April.

– 2010b. 'NDP Moves to Outlaw Genetic Discrimination.' News release, 14 April.

Newman, Jacquetta, and Linda A. White. 2006. *Women, Politics, and Public Policy: The Political Struggles of Canadian Women*. Toronto: Oxford University Press.

Nisker, Jeff, Francois Baylis, Isabel Karpin, Carolyn McLeod, and Roxanne Mykitiuk, eds. 2010. *The 'Healthy' Embryo: Social, Biomedical, Legal and Philosophical Perspectives.* Cambridge, UK: Cambridge University Press.

Nordman, A. 2004. *Converging Technologies: Shaping the Future of European Societies.* http://www.ntnu.no/2020/pdf/final_report_en.pdf.

Novas, C., and N. Rose. 2000. 'Genetic Risk and the Birth of the Somatic Individual.' *Economy and Society* 29 (4): 485–513.

Nuffield Foundation. 1999. *Genetically Modified Crops: The Social and Ethical Issues.* London: Nuffield Council on Bioethics.

Nye, David E. 2006. *Technology Matters: Questions to Live With.* Cambridge MA: MIT Press.

Organization for Economic Cooperation and Development. 2005. *The Bioeconomy in 2030: A Policy Agenda.* Proposal for a major Futures Programme Project. Paris: OECD.

– 2006. *Guidelines for the Licensing of Genetic Inventions.* Paris: OECD.

– 2010. *Statistical Definition of Biotechnology.* Paris: OECD.

O'Riordan, Timothy, and James Cameron. 1994. *Interpreting the Precautionary Principle.* London: Earthscan.

Pal, Leslie A. 2006. *Beyond Policy Analysis.* 3rd ed. Scarborough, ON: Nelson Canada, 2006.

Peekhaus, Wilhelm. 2010. 'Resistance Is Fertile: Canadian Struggles on the Biocommon.' Unpublished.

Personalized Medicine Coalition. 2009. *The Case for Personalized Medicine.* Washington, DC: Personalized Medicine Coalition.

Peters, Guy. 1999. *Neo-Institutional Theory.* London: Pinter.

Petersman, E.U., and G. Marceau. 1997. 'The GATT/WTO Dispute Settlement System: International Law, International Organization and Dispute Settlement.' *Journal of World Trade* 31 (3): 169–79.

Phillips, Peter W.B. 2002. 'Biotechnology in the Global Agri-Food System.' *Trends in Biotechnology* 9: 376–81.

– 2007. *Governing Transformative Technological Innovation: Who's in Charge?* London: Edward Elgar.

Phillips, Peter W.B., and David Castle. 2010. 'Science and Technology Spending: Still No Viable Federal Innovation Agenda.' In *How Ottawa Spends 2010–2011: Recession, Realignment and the New Deficit Era,* ed. Bruce Doern and Chris Stoney, 168–86. Montreal and Kingston: McGill-Queen's University Press.

Phillips, Peter W.B., and Robert Wolfe, eds. 2001. *Governing Food: Science, Safety and Trade.* Montreal and Kingston: School of Policy Studies, McGill-Queen's University Press.

Phillips, Peter. 2010. 'A Response to the Nuffield Council on Bioethics Consultation Paper: New Approaches to Biofuels.' Saskatoon: University of Saskatchewan.

Pollack, Andrew. 2010a. 'Awaiting the Genome Payoff.' *New York Times*, 14 June. http://www.nytimes.com/2010/06/15/business/15genome.html.

– 2010b. 'Consumers Slow to Embrace the Age of Genomics.' *New York Times*, 19 March.

– 2010c. 'FDA Faults Companies on Unapproved Genetic Tests.' *New York Times*, 11 June. http://www.nytimes.com/2010/06/12/health/12genome.html?hpw.

– 2010d. 'Genetically Altered Salmon Get Closer to the Table.' *New York Times*, 25 June. http://www.nytimes.com/2010/06/26/business/26salmon.html.

Polk, Wagner. 2003. 'Information Wants to Be Free: Intellectual Property and the Mythologies of Control.' *Columbia Law Review* 103 (4): 995–1034.

President's Council of Advisors on Science and Technology. 2008. *Priorities for Personalized Medicine.* Washington, DC: President's Council of Advisors on Science and Technology.

Prince, Michael J. 1999. 'Civic Regulation: Regulating Citizenship, Morality, Social Order and the Welfare State.' In Doern, Hill, Prince, and Schultz, 201–27.

– 2000. 'The Canadian Food Inspection Agency: Modernizing Science-Based Regulation.' In Doern and Reed, 208–33.

– 2007. 'Soft Craft, Hard Choices, Altered Context: Reflections on Twenty-Five Years of Policy Advice in Canada.' In *Policy Analysis in Canada: The State of the Art*, ed. Laurent Dobuzinskis, Michael Howlett, and David Laycock, 163–85. Toronto: University of Toronto Press.

– 2009. *Absent Citizens: Disability Politics and Policy in Canada.* Toronto: University of Toronto Press.

– 2010. 'Self-Regulation, Exhortation, and Symbolic Politics: Gently Coercive Governing?' In *Policy: From Ideas to Implementation – A Book in Honour of Professor G. Bruce Doern*, ed. G. Toner, L.A. Pal, and M.J. Prince, 77–108. Montreal and Kingston: McGill-Queen's University Press.

Purdue, D. 1995. 'Hegemonic Trips: World Trade, Intellectual Property and Biodiversity.' *Environmental Politics* 4 (1): 88–107.

Radaelli, Claudio. 2006. 'Whither "Better Regulation" for the Lisbon Agenda.' Paper presented to the ESRC Conference on EU Governance after Lisbon. Europa Institute, University of Edinburgh, 28 April.

Radaelli, Claudio, and F. DeFrancesco. 2006. *Regulatory Quality in Europe: Concepts, Measures, and Policy Processes.* Manchester: Manchester University Press.

Rath, Amitav. 2004. 'Biotechnology, Millennium Development Goals and Canada.' Paper prepared for the Canadian Biotechnology Secretariat.

Ravensbergen, Jan. 2010. 'Quebec to Pay for Fertility Treatments.' CanWest News Service, 12 March.

Raz, A.E. 2009. 'Eugenic Utopias/Dystopias, Reprogenetics, and Community Genetics.' *Sociology of Health & Illness* 31 (4): 602–16.

Reiss, Michael J., and Roger Straughan. 1996. *Improving Nature? The Science and Ethics of Genetic Engineering.* Cambridge: Cambridge University Press.

Rhodes, R.A.W. 1997. *Understanding Governance.* London: Open University Press.

Ries, Nola M., and Edna Einsiedel. 2010. *Online Direct-to-Consumer Genetic Testing: Issues and Options.* Policy brief no. 3. Ottawa: Genome Canada.

Rifkin, Jeremy. 1998. *The Biotechnology Century.* New York: Tarcher/ Putnam Books.

Royal Commission on New Reproductive Technologies. 1993. *Proceed with Care: Final Report.* Ottawa: Minister of Government Services Canada.

Royal Society of Canada. 2001. *Elements of Precaution: Recommendations for the Regulation of Food Biotechnology in Canada.* Ottawa: Royal Society of Canada.

Sample, Ian. 2010. 'Genomes of Entire Family Sequenced in World First.' *Guardian*, 10 March.

Sampogna, Christina. 2007. 'The Licensing Guidelines Project and Biotechnology IP Initiatives at the OECD.' In Health Canada, *Human Genetics Licensing Symposium*, 17–20. Ottawa: Health Canada.

Saner, Marc A. 2002. 'An Ethical Analysis of the Precautionary Principle.' *International Journal of Biotechnology* 4 (1): 81–95.

– 2010. *A Primer on Scientific Risk Assessment at Health Canada.* Ottawa: Health Canada.

Savoie, Donald. 2010. *Power: Where Is It?* Montreal and Kingston: McGill-Queen's University Press.

Scala, F. 2007. 'Scientists, Government, and "Boundary Work": The Case of Reproductive Technologies and Genetic Engineering.' In *Critical Policy Studies*, ed. M. Orsini and M. Smith, 211–31. Vancouver: University of British Columbia Press.

Schultz, Richard, and Alan Alexandroff. 1985. *Economic Regulation and the Federal System.* Toronto: University of Toronto Press.

Sell, Susan K. 1998. *Power and Ideas: North–South Politics of Intellectual Property and Antitrust.* New York: State University of New York Press.

– 2010. 'The Rise and Rule of a Trade-Based Strategy: Historical Institutionalism and the International Regulation of Intellectual Property.' *Review of International Political Economy* 17 (4): 762–90.

Shakespeare, T. 1998. 'Choices and Rights: Eugenics, Genetics and Disability Equality.' *Disability & Society* 13 (5): 665–81.
– 2005. 'Disability, Genetics and Global Justice.' *Social Policy and Society* 4 (1): 87–95.
Sharaput, Markus. 2002. 'Biotechnology Policy in Canada: The Broadening Scope of Innovation.' In Doern 2002b, 151–75.
Shiva, Vandana. 1997. *Biopiracy: The Plunder of Nature and Knowledge.* Boston: South End.
Shiva, V., and I. Moser, eds. 1995. *Biopolitics: A Feminist and Ecological Reader on Biotechnology.* London: Zed Books.
Shreeve, James. 2004. *The Genome War.* New York: Knopf.
Skinner, Brett, Mark Rovere, and Courtney Glen. 2007. *Access Delayed, Access Denied: Waiting for Medicines in Canada.* Vancouver: Fraser Institute.
Skogstad, Grace. 2008. *Internationalization and Canadian Agriculture: Policy and Governing Paradigms.* Toronto: University of Toronto Press.
Smith, John E. 2009. *Biotechnology.* 5th ed. Cambridge, UK: Cambridge University Press.
Smith, William, and Janet Halliwell. 1999. *Principles and Practices for Using Scientific Advice in Governmental Decision Making: International Best Practices.* Report to the S&T Strategy Directorate, Industry Canada.
Smythe, Stuart J. 2010. *Innovation and Liability in Biotechnology.* London: Edward Elgar.
Social Sciences and Humanities Research Council of Canada. 2010. 'Biotech Projects/Grants Data.' Ottawa: Social Sciences and Humanities Research Council of Canada.
Sparrow, Malcolm K. 2000. *The Regulatory Craft.* Washington: Brookings Institution.
– 2008. *The Character of Harms: Operational Challenges in Control.* Cambridge, UK: Cambridge University Press.
Stein, Rob. 2010. 'Company Plans to Sell Genetic Testing Kit at Drugstore.' *Washington Post,* 11 May.
Stem Cell Network. 2011. *About Us.* http://www.stemcellnetwork.ca/index.php?page=about-us&hl=eng.
Stirling, Andrew. 1999. *On Science and Precaution in the Management of Technological Risk.* Brighton, UK: Science Policy Research Unit, University of Sussex.
Sulston, John, and Georgina Ferry. 2002. *The Common Thread.* New York: Bantam.
Sutton, Sean D. 2009. *Biotechnology: Our Future as Human Beings and Citizens.* Albany: State University of New York Press.

Taylor, Charles. 1991. *The Malaise of Modernity*. Toronto: Anansi.

Taylor, Mark C. 2001. *The Moment of Complexity: Emerging Network Culture.* Chicago: University of Chicago Press.

Taylor, Paul. 2010. 'Stem Cell Therapies under the Microscope.' *Globe and Mail,* 10 June.

Thacker, Eugene. 2006. *The Global Genome: Biotechnology, Politics and Culture.* Cambridge, MA: MIT Press.

Thayler, Richard H., and Cass R. Sunstein. 2008. *Nudge.* New Haven, CT: Yale University Press.

Thompson, Grahame F. 2004. 'Is All the World a Complex Network?' *Economy and Society* 33 (3): 411–24.

Thompson, G., J. Frances, R. Levasic, and J. Mitchell. 1991. *Markets, Hierarchies and Networks: The Coordination of Social Life.* London: Sage.

Thurow, Lester. 1997. 'Needed: A New System of Intellectual Property Rights.' *Harvard Business Review* 75 (3): 94–103.

Tiedemann, Marlisa. 2006. 'Review of Bill C-5: Public Health Agency of Canada Act.' Ottawa: Parliamentary Information and Research Service, Law and Government Division, LS-523E.

Trebilcock, Michael, and Robert Howse. 1995. *The Regulation of International Trade.* London: Routledge.

Turney, Jon. 1998. *Frankenstein's Footsteps: Science, Genetics and Popular Culture.* New Haven, CT: Yale University Press.

Tyler, I. 2010. 'Designed to Fail: A Biopolitics of British Citizenship.' *Citizenship Studies* 14 (1): 61–74.

United Nations Environment Program. 2000. 'Draft Cartagena Protocol on Biosafety.' Montreal: UNEP.

U.S. Embassy. 2000. 'Byliner: Assistant Secretary of State Sandalow on Biosafety Protocol.' News release, 11 February. Geneva.

Van Tassel, Katherine. 2009. 'Genetically Modified Food, Risk Assessment and Scientific Uncertainty Principles: Does the New Understanding of the Networked Gene Trigger the Need for Post-Market Surveillance to Protect Public Health?' *Boston University Journal of Science and Technology Law* 15: 220–51.

Vogel, David. 1995. *Trading Up: Consumer and Environmental Regulation in a Global Economy.* Cambridge, MA: Harvard University Press.

– 1998. 'The Globalization of Pharmaceutical Regulation.' *Governance* 11 (1): 1–22.

Vos, E., and Govin Permanand. 2009. 'Between Health and the Market: The Roles of the European Medicines Agency and European Food Safety Authority.' In *Health Systems Governance in Europe: The Role of EU Law and*

Policy, ed. E. Mossialos, G. Permanand, and T. Harvey, 134–47. Cambridge: Cambridge University Press.

Wade, Nicholas. 2010a. 'A Decade Later, Genetic Map Yields Few New Cures.' *New York Times*, 12 June. http://www.nytimes.com/2010/06/13/health/research/13genome.html.

– 2010b. 'Researchers Say They Created a "Synthetic Cell."' *New York Times*, 20 May.

Wellcome Trust. 1999. 'From Genome to Health.' *Wellcome News* 20 (3): 10–27.

Wiles, Anne. 2007. 'Strategically Natural: Nature, Social Trust and Risk Regulation of Genetically Modified Foods and Natural Health Products in Canada.' PhD diss., Carleton University.

Williams, Garth. 2006. *Trends in North American Research on the Ethical, Legal and Social Aspects of Genomics*. Ottawa: Analysis of Discussions at SSHRC-ERA-SAGE Workshop, 22–23 May.

Williams, Glen. 2009. *Canadian Politics in the 21st Century*. 7th ed. Toronto: Nelson Canada.

Williams-Jones, T. 1999. 'Re-framing the Discussion: Commercial Genetic Testing in Canada.' *Health Law Journal* 7 (2): 49–67.

Wittman, Hannah, Annette Aurélie Desmarais, and Nettie Wiebe, eds. 2010. *Food Sovereignty: Reconnecting Food, Nature and Community*. Halifax: Fernwood.

World Commission on Environment and Development. 1987. *Our Common Future*. Oxford: Oxford University Press.

World Health Organization. 2002. *Genomics and World Health*. Report of the Advisory Committee on Health Research. Geneva: World Health Organization.

Yee, S. 2009. 'Gift without a Price Tag: Altruism in Anonymous Semen Donation.' *Human Reproduction* 24 (1): 3–13.

Index